IMAGES
of America

ST. PETERSBURG'S MARITIME SERVICE TRAINING STATION

With Albert Whitted Municipal Airport in the background, the Color Guard from U.S. Maritime Service Training Station St. Petersburg represented the proud maritime traditions of old and the necessity of military-like training and discipline needed to keep American ships moving before, during, and after World War II. (USF.)

On the Cover: Merchant seamen trainees practice lifeboat procedures off the shore from the Vinoy Hotel. Located two miles north of the training station, the hotel was used to house trainees and officer candidates after the United States entered World War II. (USF.)

IMAGES
of America

St. Petersburg's Maritime Service Training Station

Michelle L. Hoffman

ARCADIA
PUBLISHING

ISBN 978-0-7385-4250-8

Published by Arcadia Publishing
Charleston, South Carolina

Printed in the United States of America

Library of Congress Control Number: 2005938281

For all general information, please contact Arcadia Publishing:
Telephone 843-853-2070
Fax 843-853-0044
E-mail sales@arcadiapublishing.com
For customer service and orders:
Toll-Free 1-888-313-2665

Visit us on the Internet at www.arcadiapublishing.com

To Jonathon and Carmen:

Spend your life with someone who inspires your dreams

and is willing to be an unwavering companion on the road to reach them.

I hope you are blessed with someone who is as supportive and loving as your father.

CONTENTS

ACKNOWLEDGMENTS

Motivation for research like this comes from many sources and in many forms. The two organizations that I cannot thank enough are University of South Florida St. Petersburg and the American Merchant Marine Veterans Association. The impetus for this book began with the collection of photographs and documents donated by Donald Saxer, who was part of the stand-by crew when the U.S. Maritime Service Training Station (USMSTS) St. Petersburg was decommissioned in 1950. Without Saxer's initial donation to the Poynter Library, the story of 25,000 men would have been left to newspaper reports and scattered records in the National Archives. At the University of South Florida St. Petersburg, I'd like to thank Jim Schnur, special collections librarian, for his encouragement and endless hours of availability as this project evolved. In the Florida Studies Institute, Drs. Ray Arsenault and Gary Mormino challenge me on a weekly basis not just to be a good student, but a better historian and much better teacher.

From the American Merchant Marine Veterans Association, a sincere "thank you" to all of the national and chapter officers around the country who have spread the news about this project. I would like to especially thank those St. Petersburg personnel, trainees, and their families who contributed photographs and memories to this project: George Bark; William Bentley; Rex Farley; Kenneth Fountain; Chas Keller, on behalf of his grandfather Charles Keller; Edmund Kinter; Frank Olanski; Donald Saxer; and Aaron Varn. It had long been Mr. Saxer's wish to see this history documented and shared. His excitement and love of the U.S. Maritime Service Training Station inspired me to finish this project at a time when life's fatigue was overwhelming both of us. Mr. Saxer died on February 14, 2006, just days before this book went to print.

Also I would also like to extend my appreciation to the John F. Germany Library in Tampa, Florida; the Lakeland Public Library in Lakeland, Florida; the archives staff at the George A. Smathers Library at the University of Florida; Nashville Public Library in Nashville, Tennessee; and the Still Pictures Branch of the National Archives in College Park, Maryland.

KEY FOR COURTESY INFORMATION:
GB—George Baker's personal collection
USF—Special Collections Department at the University of South Florida St. Petersburg
UF—J. Hardin Peterson Collection at the University of Florida
KF—Kenneth Fountain's personal collection
MK—Martha Keller's personal collection
EK—Edmund Kinter's personal collection
AV—Aaron Varn's personal collection
DS—Donald Saxer's personal collection
RFi—Written correspondence from Rex Farley
KFi—Written correspondence from Kenneth Fountain
EKi—Written correspondence from Edmund Kinter
LOC—Library of Congress, Prints and Photographs Division, FSA-OWI Collection
NARA—National Archives Still Photographic Division
USMS—United States Maritime Service, Preliminary Training Manual
GBi—Interview with George Bark
DSi—Interview with Donald Saxer
AVi—Interview with Aaron Varn
DQi—Interview with Denis Quilligan

INTRODUCTION

Although the geography of Bayboro Harbor is small compared to other inlets in the Tampa Bay estuary, local knowledge of its history is just as contained. One such story is that of the U.S. Maritime Service Training Station and how it came to be in St. Petersburg, Florida. Nestled just south of downtown, Bayboro Harbor was not the U.S. Maritime Commission's first choice for its Gulf Coast training facility in 1939. The award for the seamen training site was actually given to the city across the bay—Tampa. Formed by Congress three years earlier, the purpose of the U.S. Maritime Service was to train merchant seamen under the administration of the Coast Guard. As a result, the selection criteria for its training facilities included position of Coast Guard facilities and access to instructors and the location's conduciveness for year-round instruction as well as its potential for recruiting efforts.

Tampa was ranked as the 11th largest port in the country in 1938 and had recently been awarded several multi-million-dollar shipbuilding contracts by the Maritime Commission. Coupled with the presence of a local navigation school and the U.S. Public Health Quarantine Station at Gadsden Point, Tampa seemed to be the perfect location.

Like many metropolitan areas of the time, Tampa hoped that by securing military and government programs the effects of the Depression would ease. The problem was that while Tampa had offered Gadsden Point to the Maritime Commission, they simultaneously had presented the rest of the Interbay Peninsula to the War Department as part of a bid for the Southeast Air Base. Unknown to local officials, military leaders hoped to include the land and facilities at Gadsden Point to ensure that the air base would encompass the majority of the peninsula. So when the War Department found out about the Maritime Commission's decision to place a training facility at Gadsden Point, it threatened to relocate the base. Tampa leaders quickly decided that the air station was much more important to the city's vitality in the long run and declined the offer for the maritime service training school.

As word got out that Tampa had declined the merchant marine training station in the latter weeks of October 1939, cities along the Gulf Coast scrambled to get their proposals in front of the Maritime Commission's leadership, and St. Petersburg was no different. Besides its comfortable climate, this popular tourism spot had several advantages from the beginning. The most obvious was the city's long relationship with the U.S. Coast Guard, the entity responsible for overseeing the training of the merchant seamen. To strengthen its position even further, the Sunshine City proposed a location for the maritime school adjacent to the Coast Guard base, allowing for easy billeting of instructional personnel and staff. But the city's secret advantage was someone within the Maritime Commission itself.

Knight Aulsbrook, a Roosevelt appointee to the Maritime Commission and son of a very prominent local real estate developer, visited St. Petersburg the day before he attended the launching of a new cargo ship, *Sea Witch*, in Tampa. In previous years, St. Petersburg officials used a "come and see" strategy in securing government facilities such as the Bay Pine Soldiers Home, expanding Coast Guard facilities and obtaining federal funding for port improvements. Although the *St. Petersburg Times* claimed that Aulsbrook's trip to St. Petersburg was merely to see his parents, the chain of events following his visit illuminated the influence this local son carried within the Maritime Commission. Within that week, St. Petersburg leaders made their official presentation to the government. On the evening of November 2, 1939, two days after the city's presentation, Knight's father contacted the *St. Petersburg Evening Independent*, leaking the news that the Maritime Commission had selected St. Petersburg as the site of the Gulf Coast maritime service training station.

. The official announcement came on November 6 as city leaders began making preparation for the arrival of two training ships, SS *American Seaman* and *Joseph Conrad*. Two customhouse buildings were relocated to the seven-acre tract where the maritime service training station would be located. These buildings acted as temporary offices for clerical and planning staffs while the plans were formalized for the construction of base facilities.

It is important to understand that the purpose of the maritime service was to train better men for the merchant marines of the United States. Although the necessity of maritime vessels and crews in times of war were and still are vital to the success of any military operation, the U.S. Maritime Service was not a war-time organization and was purely voluntary. But this reality became blurred with the onset of World War II and transfer of the Coast Guard to the Navy Department.

A three-masted full rigger, the *Joseph Conrad*, was built in 1882 in Denmark and had been used for 52 years as a training ship for the Danes. In 1939, the ship's ownership was transferred to the Maritime Commission, where it was used to apprentice seamen the fundamentals of deck seamanship. In 1945, the 111-foot vessel was retired and became the property of Mystic Seaport. Today it is used as an exhibit and training ship for a youth sailing camp in Mystic, Connecticut. (USF.)

One

CONSTRUCTION AND FACILITIES

The initial construction of facilities at USMSTS St. Petersburg included a two-story, T-shaped, solid concrete building for dormitories and staff quarters, galley, auditorium, library, and administrative offices. In 1940, the federal government outlaid $750,000 in initial building costs. (DS.)

With confirmation of the city's winning of the maritime service training station, officials immediately began construction as promised. And with *Tusitala* already moored at the dock and the bulkheads completed, the foundation for the main building was ready to be laid. As seen in the photograph, William Knaver's company provided the ready-mix concrete for the building. (USF.)

Still reeling from the later years of the Great Depression, the construction of the base brought much needed business to the city and the rest of Central Florida. St. Petersburg–based subcontractors included the Pinellas Lumber Company, the hardware store Cunningham Brothers, plumber Lee Sheffield, local painter W. F. White, floor sander Nihls Hallgren, and mason James Riddle. (USF.)

From January 1940 until the main building construction was completed in July 1941, approximately 900 men satisfied the training courses and were released for employment on U.S. merchant vessels. (USF.)

As the headquarters for the training station, this facility housed over 250 trainees and officers. This view of the building shows the seawall or bulkhead built by the city. With construction completed in the summer of 1941, training ships were moored along this wall until the base's decommissioning in 1950. (USF.)

Trainee George Bark had enlisted in the maritime service at Hoffman Island, located just outside of New York harbor in the late summer months of 1941. Upon arrival in St. Petersburg, he remembers standing outside for muster after being issued wool, dress-blue uniforms and watching other guys faint from the Florida heat: "We weren't use to the heat." To help contain construction cost, seamen trainees were used to finish the landscaping around the Administration Building after their instructional classes were completed on the *American Seaman* and *Joseph Conrad*. (GB.)

This aerial view revealed completed landscaping and the cleared area where barracks, pool, medical facilities, and a physical fitness arena were to be built by 1943. (NARA.)

With the *Joseph Conrad* as a perfect backdrop, graduation in spring 1942 highlighted the successful completion of the six-month training program. During this first year of training operations, the average age of a seaman apprentice at USMSTS St. Petersburg was 20 years old. (USF.)

Inspections were held on a regular basis in front of the Administration Building until the parade grounds, which faced First Avenue, were completed. The original flagpole still stands today, and the building is now used as the College of Marine Science at the University of South Florida St. Petersburg. (USF.)

With the training vessel *Tusitala* in the background, the main entrance to the Administration Building greeted visitors with two brass cannons from the *Joseph Conrad* aimed toward the flag pole. Donald Saxer recalled, "When visitors came in that door, if they turned left they saw the doors leading to the sick bay. If they turned right, there were administrative offices for the civilian workers, the executive officer and the captain." (USF, DSi.)

During the hot summer months of 1941, George Bark remembered being on guard duty in a makeshift shack with boards holding up the sides and construction debris along the fence line. "It's just another example of how unready the place was." (GB.)

Within a year, the guard shack was transformed into a true main gate. Hundreds of men passed through it daily on their way to and from liberty. It was also the first glimpse of the base for trainees upon their arrival. (USF.)

On January 30, 1943, the second phase of construction was completed with the dedication ceremonies for the barracks and mess halls as well as for the academic and infirmary buildings. The cost of the additional facilities was approximately $325,000. Present at the ceremonies were Adm. Telfair Knight, director of the Training Division for the Maritime Commission; Capt. P. C. Mahady, superintendent of the training station; Comdr. F. R. Francke, Naval Reserve intelligence officer and local St. Petersburg resident; and Mayor Robert McCutcheon of St. Petersburg. (USF.)

The barracks buildings contained eight dormitories. Each dormitory was capable of accommodating approximately 100 men. In addition, this facility contained two large recreation rooms on the second floor. (USF.)

In 1943, an expansion program added two mess halls. With one galley or kitchen connecting both dining areas, 600 enrollees could eat meals at the same time. To expedite food service, meals were served in a buffet style using two steam tables in each hall. (USF.)

Everything from soups to vegetables to meats were prepared in these huge steam coppers. The galley also contained six ovens as well as four walk-in freezers. (NARA.)

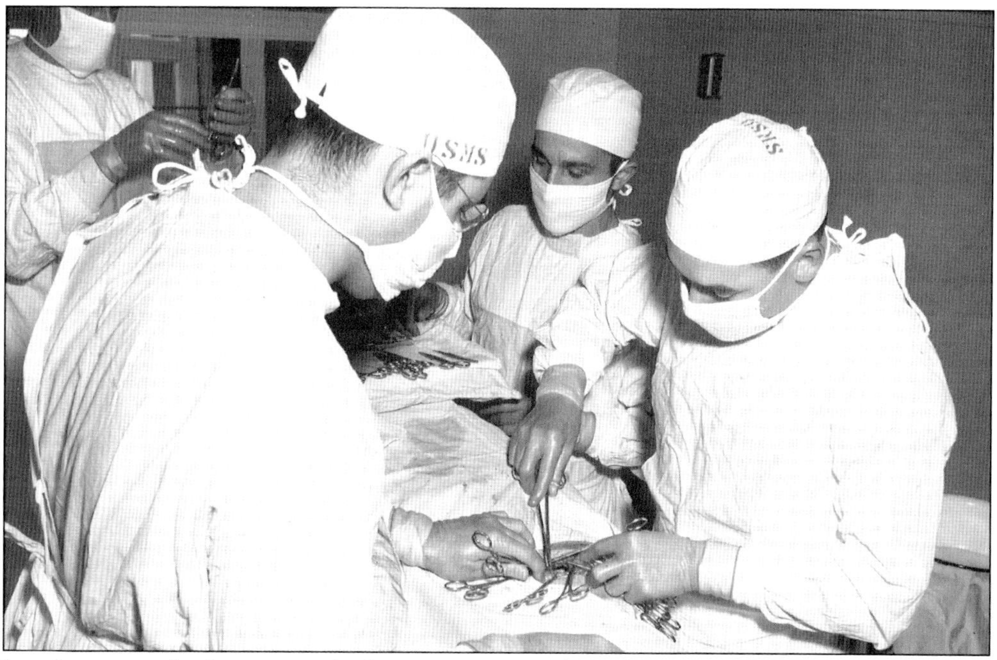

In order to meet the dynamic needs of a growing training base, the Maritime Commission authorized a full medical staff to be located at USMSTS St. Petersburg, comprised of doctors and nurses from the U.S. Public Health Service and with the maritime service personnel in supplemental roles. The infirmary was a two-story framed structure that contained five wards, a diet kitchen, and a processing room where enrollees were given their physical examinations. (USF.)

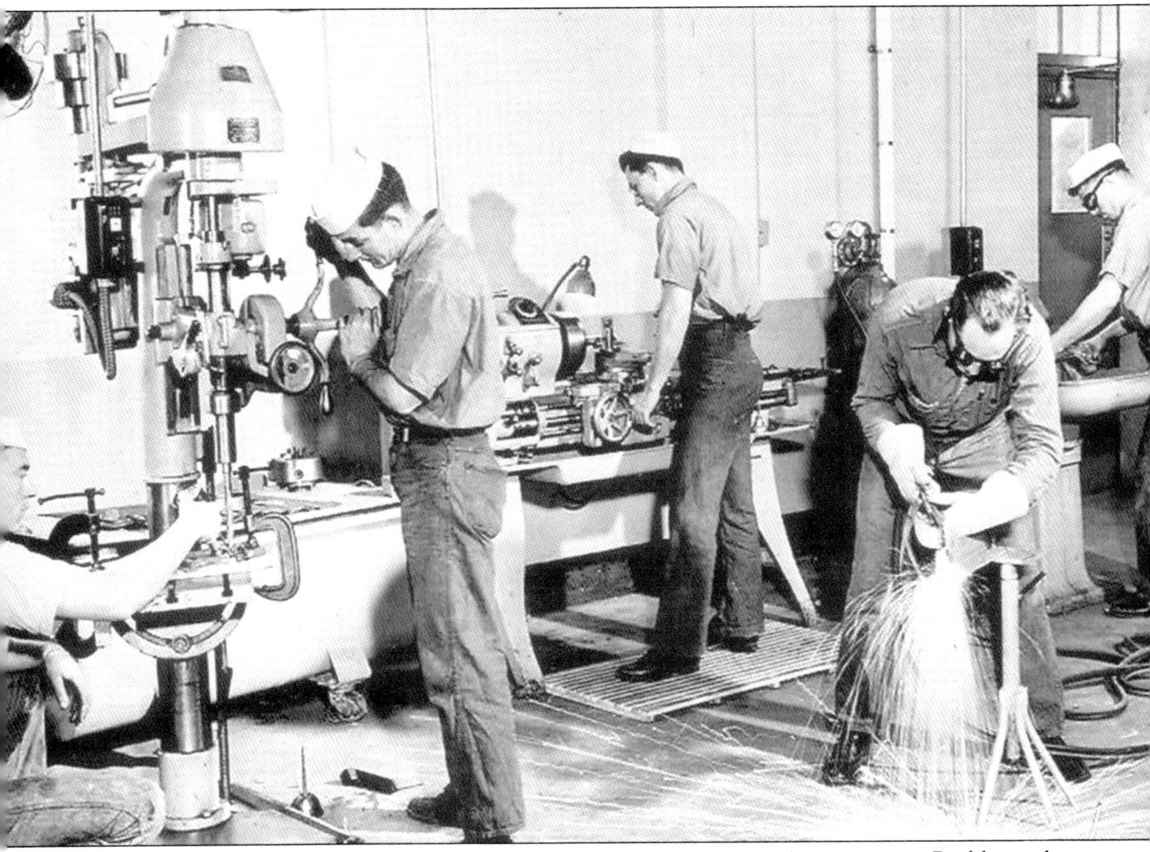

Located next to the wooden boat docks on the west side of the Administration Building, the Academic Building housed this machine shop, the clothing supply and induction room, the ship's service store, the post office, storerooms, and maintenance offices on the first floor. On the second floor, there were 10 large classrooms and offices for instructors and training personnel. (USF.)

With slips for 18 rowboats and 2 stationary docks, the boat dock was along the western wall of the peninsula. This seawall was constructed with concrete unlike the east bulkhead, which was made of steel. (USF.)

Kenneth Fountain remembered how important boxing was to the trainees. "Every Friday night, we were treated to boxing matches," said Fountain. "The winners were given the whole weekend off. My best buddy tried to get me to fight him on the regular card, which was okay with me, but he wanted me to throw the fight so he could get the weekend off. Now I was an innocent boy from a small town, but I was not dishonest. So I turned him down." (KFi.)

Behind the barracks on the north side of the base, an outdoor athletic ring was built for boxing and other exhibition sports. With a light box hanging over the ring that read "Get in the fight," trainees were encouraged to participate in boxing matches. (USF.)

Former trainee Donald Saxer recalled the fights quite well: "One guy who did so much for the boxing matches was Charles Warren. After his father died when he was teenager, he fought back in Tennessee for $5 to $10 against older men. He had his nose broken so many times his boxing nickname became 'The Great Goose.' He was something else." (DSi.)

According to the *Nashville Tennessean*, Warren won the Florida heavyweight Golden Gloves championship in 1945 while he was stationed at the maritime service training station. (DS.)

After his service in St. Petersburg, Warren later competed for the heavyweight championship for the Mid-State Golden Gloves at the Hippodrome on West End Avenue in Nashville, Tennessee. According to the local newspaper report, he lost the January 26, 1949, match to Lillard Gibbs, a local Nashville athlete from Tennessee Tech. (DS.)

The base also had an auditorium complete with a stage for award ceremonies, a projector screen, and a state-of-the-art film projector. With a seating arrangement of 38 rows of 11 plus an additional 19 rows of 6, the room was large enough to seat more than 500 personnel. (USF.)

Even after the war began, Tuesday night dances were held in the auditorium. A group of local young women known as the Bomb-a-Dears participated in every station dance since its inception. (USF.)

"If a ship was sinking, the merchant seamen would have to abandon ship using ropes," said Donald Saxer. "If rescued, they might have to climb up a rope to get aboard. So the purpose of the rope-climbing apparatus was to get the men familiar and comfortable using the ropes." (DSi.)

The Olympic-size swimming pool ensured that trainees knew how to swim and were able to perform a variety of procedures, including how to enter a boat from the water and how to haul a man into the boat without tipping over. Trainees also treaded water for extended periods of time to guarantee they could use an assortment of floating techniques when fatigued. (DS.)

Located at the deep end of the training pool, the climbing platform allowed trainees to practice the proper method for safely abandoning ship. There were also different methods for floating with and without life vests as well as methods for how to swim safely through blazing oil on the surface of the sea. Trainees also learned how to abandon ship using a type of life vest called the "cork jacket." To prevent the hard cork jacket from riding violently up the body when hitting the water and possibly delivering a knock-out blow to the chin, a trainee practiced holding down the top of the jacket with his arms held close to the sides. (USF.)

Two

Enrollees and Indoctrination

The maritime service wasted no opportunity to recruit seamen between the ages of 17-and-a-half and 52. Prior to the attack on Pearl Harbor in 1941, 19-year-old Aaron Varn, a resident of Durant, Florida, became interested in the merchant marines after hearing a radio commercial on WSUN. He said, "So I sent them a penny postcard asking for an application. About a week later the application came, and I mailed it back. And in about week after that I got a letter telling me to report in a week." (NARA, AVi.)

During the war years, recruiting initiatives led by Lieutenant (junior-grade) Hubbard took advantage of Florida's maritime industries by setting up displays in prominent locations and speaking with men of all ages and backgrounds. (DS.)

Rabbi Samuel Rubin enrolled in the U.S. Maritime Service in November 1944. The oath of allegiance recited by the seamen recruits paralleled that of the military branches, but what made the maritime service different from the military was that trainees could disenroll at their request. Former trainees were responsible for securing their own transportation back home. (NARA.)

Already two years at sea when this photograph was taken in 1944 by the Division of the Public Relations, 22-year-old Ted Soteroplos commented, "My brother is in the army in the South Pacific. . . . Now more than ever our army needs the planes, tanks, and guns. Our job is to see that they get them." (NARA.)

Standing in the middle of his recruits, Chief Donald Saxer traveled with these young men from the Tampa Bay area to the Sheepshead Bay Training Station in Brooklyn, New York. When they left Tampa, it was 65 degrees. Upon their arrival in New York, it was 10 degrees and snowing. (DS.)

Seamen recruits arrived in St. Petersburg from around the country via the railroad. The *Chicago Sun* referred to USMSTS St. Petersburg as "a place where a man who had never set foot on a ship would learn within a week's time what it meant to 'jibe the spanker' or 'stretch the outhawl.' " (USF.)

Upon arrival at the base, trainees presented their papers and travel orders to the quartermaster at the front gate. Ed Kinter wrote, "We were greeted by recent recruits jeering, 'You will be sorry. Just wait till you get those shots. They use doll needles with barbs on them.' " (USF, EKi.)

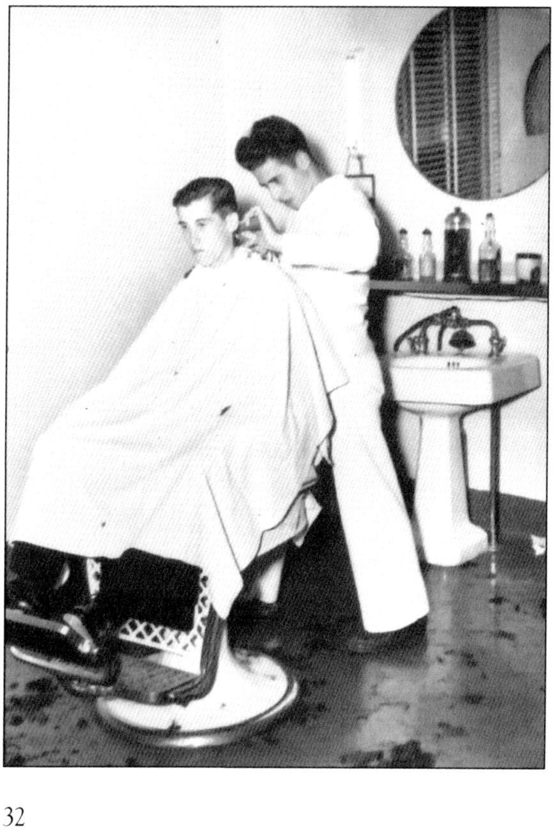

Once received by the quartermaster, trainees filed into the barber's chair for the beginning of their transformation into seamen trainees. (NARA.)

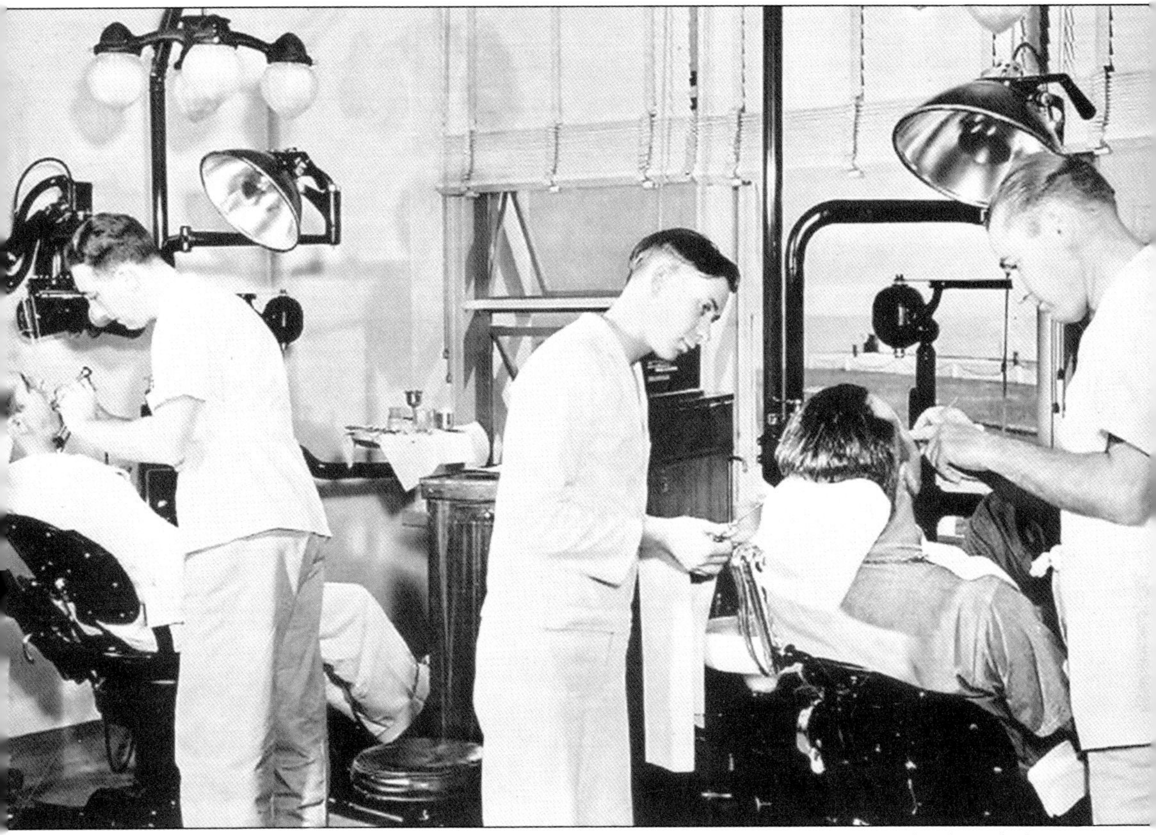

Every man was given a thorough dental examination upon entering the station. The dental clinic had five chairs in operation continuously. Dr. J. Walton Lawrence was the resident dental surgeon with Robert J. Herder serving as the assistant dental surgeon at USMSTS St. Petersburg. Both men were part of the personnel from the U.S. Public Health Service assigned to the training station. (LOC.)

With haircuts and medical examinations completed, the next hour of these trainees' first day at the maritime base was one most never forgot. First each was measured from head to toe and then filed into the undressing room, where he removed the last shred of his civilian life. All of his civilian clothes were then placed in wrapping paper, tied with cord, and sent home. From there, the trainee "divorced himself of his civilian germs" by using a bar of special germicidal soap. While his feet soaked in a chemical bath, the recruit received a handbag containing two towels, four pairs of socks, four pairs of underwear drawers, a pair of black shoes, and tennis shoes. According the *Maritime Service News*, the trainee moved up to the counter, where he received two pairs of dungaree trousers, four cotton shirts, four chambray shirts, a jersey, a pair of bathing trucks, a pea jacket, and other sundry articles. The approximate value of this first issue was $40 per man. (USF.)

Ed Kinter enlisted in the maritime service on August 1, 1943, in Pittsburgh, Pennsylvania. He remembers hearing an indoctrination speech after receiving his uniform. According to base documents, the speech likely emphasized recognition and responsibility: "Before entering the U.S. Maritime Service, you were responsible only to yourself. In the uniform of the Maritime Service, you are a representative member of the American Merchant Marine. . . . We have a glorious tradition, written in the blood and bravery of thousands of our fellow seamen. To conduct ourselves in a manner worthy of this tradition is our responsibility." (EKi.)

After the indoctrination speech, an identification card was issued by the captain of the Port of St. Petersburg to each seamen trainee. (EK.)

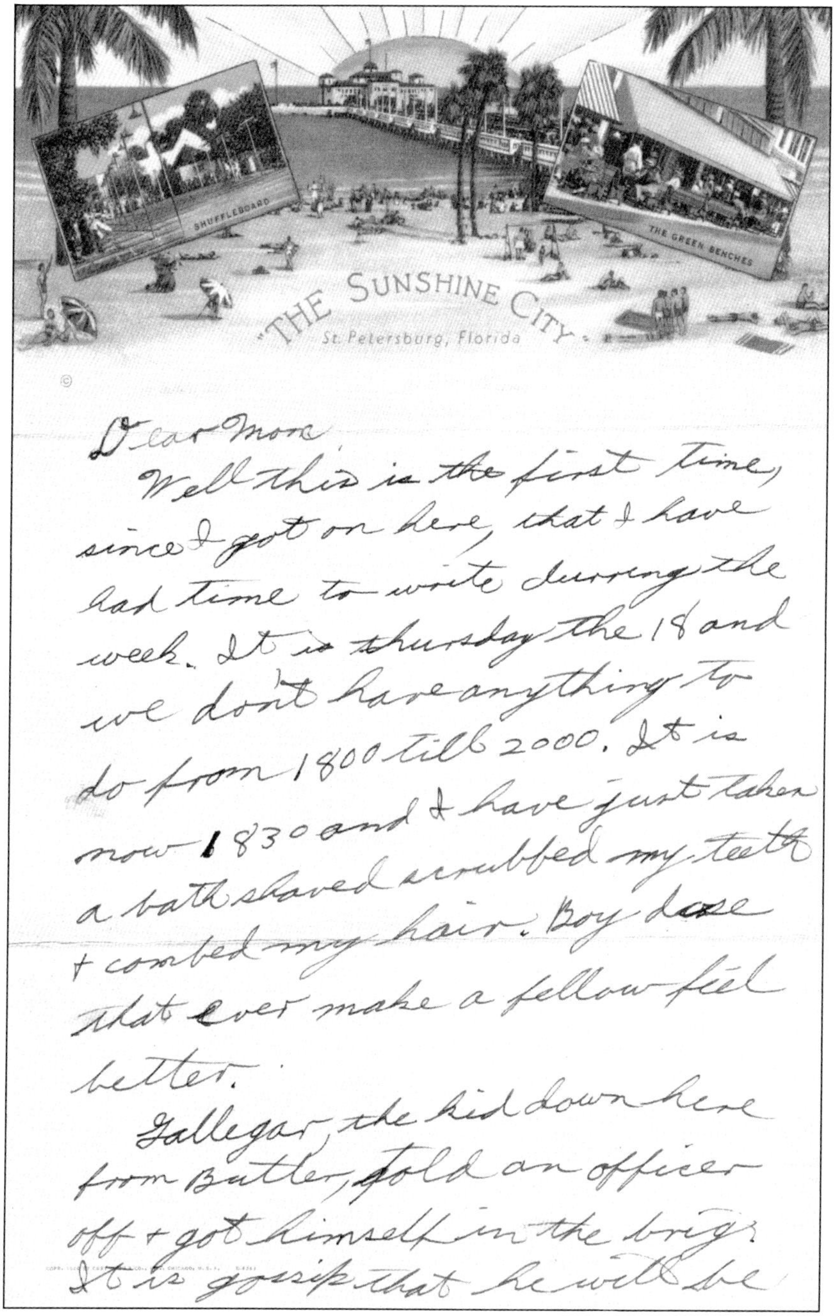

Dear Mom
Well this is the first time since I got on here, that I have had time to write during the week. It is thursday the 18 and we dont have anything to do from 1800 till 2000. It is now 1830 and I have just taken a bath shaved scrubbed my teeth + combed my hair. Boy dose that ever make a fellow feel better.
Gallegar, the kid down here from Butler, told an officer off + got himself in the brig. It is gossip that he will be

This letter from Edmund Kinter recounts how busy trainees were upon their arrival. For Kinter, the trip to Florida was most memorable after a Pullman porter returned his money belt, which Kinter had stored in a pillowcase for safekeeping. "I believe this was my first large step toward trusting people." After the maritime service, Kinter, who trained in the cooks and bakers program, entered the food service industry, where he worked as a restaurant manager for the G. C. Murphy Company, and was married in 1951. In 1971, he became a district manager with oversight of 20 locations and was promoted to director of food service 10 years later. In 1989, he retired from the food industry and today spends his days playing golf, gardening, and raising honeybees. (EKi.)

Three

Preliminary Training

A trainee's first five to eight weeks of instruction were spent mostly on shore. Among the many topics covered, a trainee received significant instruction in fundamental seamanship, became familiar with nautical terms, practiced the procedures for making knots and splices, and learned how to properly row and command a lifeboat. (USF.)

The maritime service assumed that the majority of men training in St. Petersburg had never been to sea. As a result, the *Training Manual* began with the following introduction: "It is assumed that you will have a good understanding of nautical terminology and that you will not embarrass yourself or your shipmates by carrying around with the appendages of the landlubber." (USMS.)

The old system of signaling on the ship's bell to the mate on the bridge was one of the traditions learned by seamen apprentices during the first eight weeks. The bell system is as follows: one bell, object on the starboard (right) side; two bells, object on the port (left) side; three bells, object dead ahead. (USMS.)

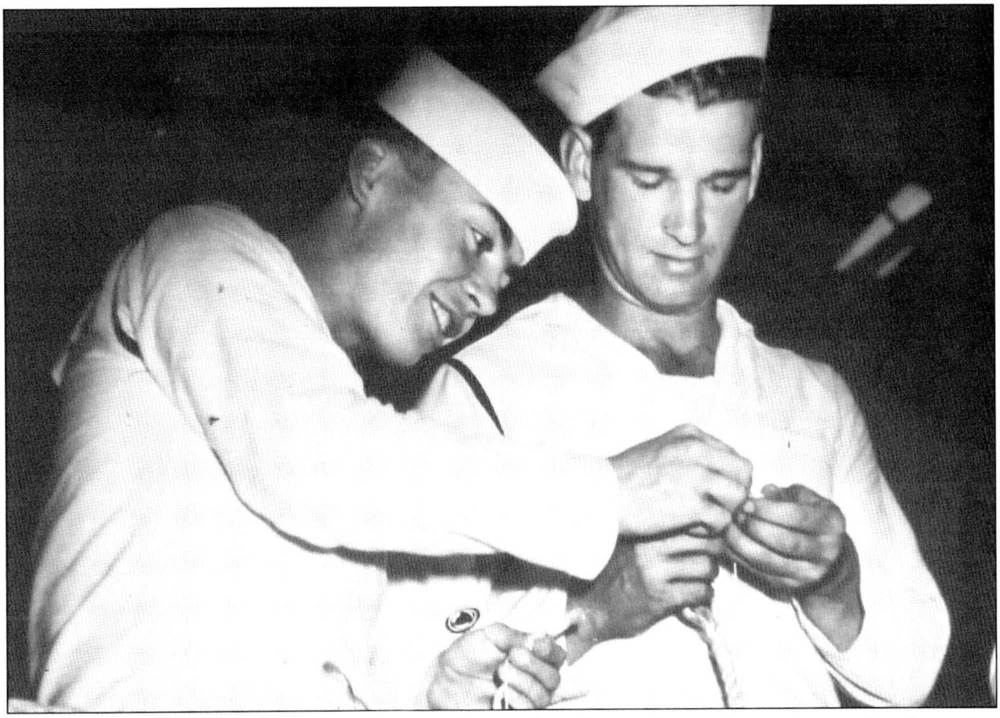

The *United States Maritime Service Training Manual* emphasized to trainees the need to "practice these knots with a small length of line. Do not get in the habit of hand tying; tie knots in awkward positions as you may aboard ship. Put them to practical use." (USMS.)

Trainees learned that "the safety of a vessel and its crew frequently depends on simple knots used in joining lines together. For this reason, no man can consider himself a seaman until he has mastered the methods of quickly and properly tying these common used knots," stated the *American Merchant Seaman's Manual.* (USMS.)

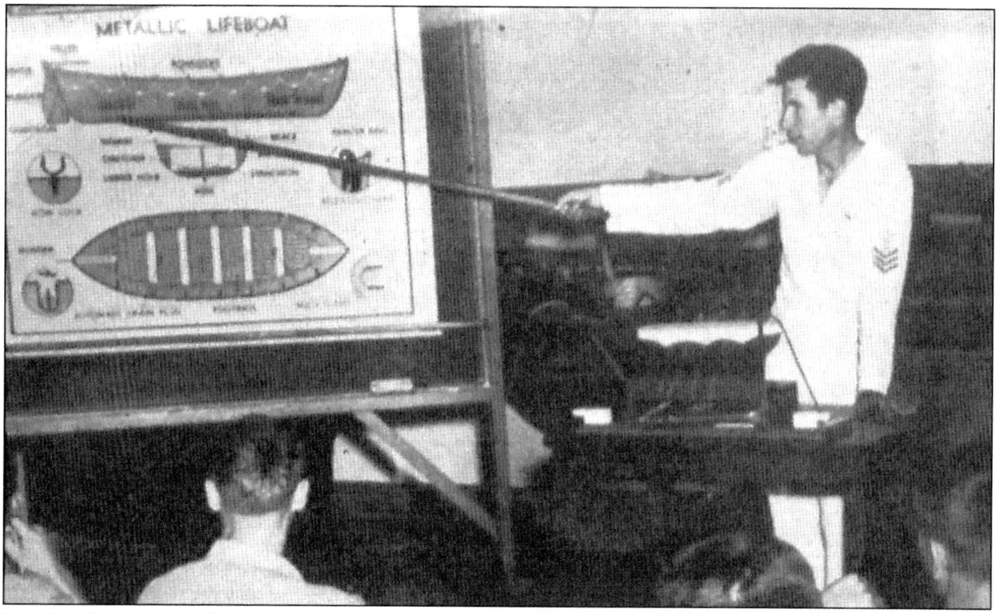

In a brochure titled "St. Petersburg's Training Men," the necessity of understanding all parts of the lifeboat were explained: "In almost every marine disaster where the lifeboats are used, casualties result because people fail to understand some of the simple things about boats. Every man who goes to sea must learn these things so that he may some day be able to assist in saving the lives of others as well as his own." (USMS.)

"Lifeboat davits are of many types with slight differences in each," stated the *United States Maritime Service Training Manual.* "The most common are the gravity, radial or round-bar and quadrantal. The Liberty ship davit might be considered a variation of the old type of quadrantal davit. On most vessels, the lifeboats are now secured over the side by pelican hooks for immediate release." The davit is the crane used to lower and lift lifeboats. The davit could be manually powered as seen in the photograph or powered through hydrolics. (USMS.)

The *United States Maritime Service Training Manual* provided detailed descriptions about how oars operate: "The action of the oars on either side is like that of a paddle wheel steamer, and going ahead on one side and backing on the other will turn the boat rapidly upon her pivoting point. The boat commands are like the telegraph to the engine room. If the coxswain cannot tell the oarsmen what to do, he cannot expect to get it done, and if the oarsmen don't understand the commands, they won't get much done either." (NARA.)

As if it were having a dialogue with the trainee, the *United States Maritime Service Training Manual* explained the importance of practicing what was just learned: "Having learned the commands, we are in a position to consider the art of rowing, for it is an art, and like other arts, is only learned by practicing. The whole secret of pulling an oar is the body swing and good leg work against the stretcher." To ensure mastery of rowing, the training station held regularly scheduled boat races in Bayboro Harbor regardless of weather conditions. (NARA.)

According to training materials, "The object of sail training is to teach enough of the fundamentals of ocean sailing, so that you will be of worthwhile assistance in sailing your lifeboat to safety, in event that your ship is torpedoed far from land." (LOC.)

"Seamanship on a small boat off-shore, particularly an open boat, is largely a matter of constant vigilance and good judgment displaying itself as caution. In fact, the more experienced you are the more vigilant and careful you are apt to be," warned the *American Merchant Seaman's Manual.* "When you are a thousand miles from land, you don't have a 'second chance.' " (NARA.)

Instruction in the proper wearing of a life jacket could not be over-emphasized. All trainees also received thorough training, both theoretical and practical, in other types of lifesaving equipment, such as life rings, water lights, exposure suits, safety bags, and kits. Effective learning of this phase was achieved through constant drills. (USF.)

Abandon-ship drills prepared the future merchant seaman to handle himself intelligently in the event of an emergency. Those merchant seamen who were forced to abandon ship by reason of enemy attack strongly urged that repeated lifeboat drills be conducted. (LOC.)

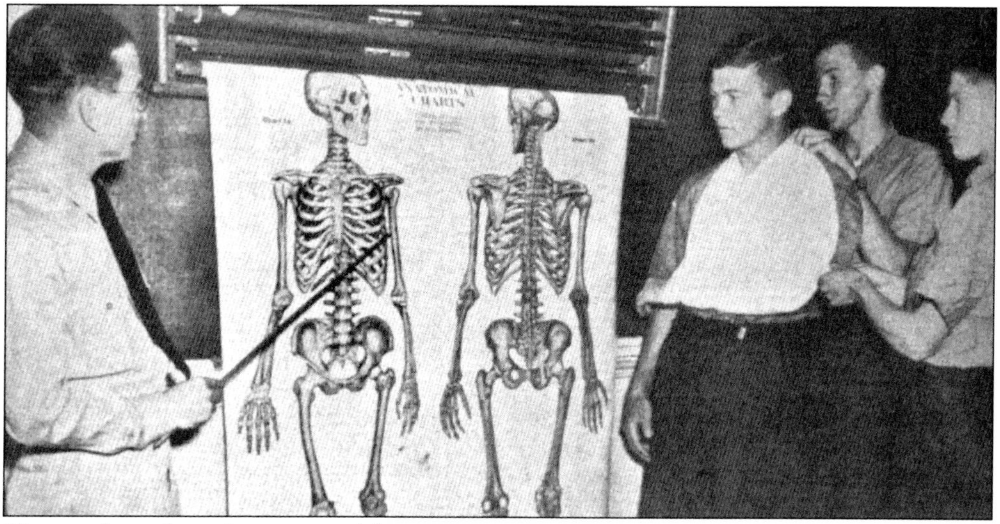

Wartime hazards made it essential for every trainee to possess at least a basic knowledge that enabled proper application of first aid. This instruction included procedures for the treatment of burns, application of tourniquets, coping with different types of unconsciousness, and methods of resuscitation. (USMS.)

Also during the preliminary training, seamen apprentices were taught the basics of firefighting as well as how to use the various types of fire extinguishers and other equipment essential to combating the assortment of fires that might occur aboard ship. (USF.)

Trainees also had to learn the types of fire. Class A fires, or general fires, were the most common and generally occurred where the crew lived, ate, and worked. This type of fire required the use of a soda acid extinguisher. In the engine room where everything was metal, the danger of fire came primarily from oil or other types of inflammable liquids of the oil family. As a result, the foam-type extinguisher was used. (DS.)

Trainees very quickly learned what the *United States Maritime Service Training Manual* meant when it stated, "Fire is the union of any substance with oxygen. To destroy fire, it must be cooled or the oxygen supply cut off or the combustible material removed." (USMS.)

At the maritime service Fire Fighting School, located about eight miles north of the training station near an area known as Toytown, trainees learned how to prepare the hoses while dealing with smoke and heat. (DS.)

As trainees battled the flames, they remembered what they read in the *Training Manual*: "Anything will burn but if caught in time and skillfully handled many fires on shipboard that otherwise might be dangerous, can be extinguished." (USMS.)

For trainees like George Bark, shown on the far left, who arrived in St. Petersburg before the completion of the base's buildings, preliminary training occurred on ship as well as on land. The world's smallest full-rigger, *Joseph Conrad*, had spent its entire existence as a training vessel for the Danish, the British, and now the American Merchant Marines. (GBi.)

Seamen apprentices went from practicing abandon-ship drills on the seawall to practicing them from the *Joseph Conrad* while it was underway. (LOC.)

The physical exertion of the rope-climbing apparatus paid off for most trainees when they transferred to the *Conrad* and were told to "man the yards." Aaron Varn recalled that it was better to be at the top of the mast when the sails were unfurled "because if it had rained, you got really wet when the sails opened." (LOC, AVi.)

The tropical weather of coastal Florida forced the St. Petersburg trainees to adapt to working in sudden downpours, especially while on the *Joseph Conrad*. (NARA.)

Trainees were sometimes assigned to "holystone" the deck of the ship. As a method of cleaning the wooden floors, the first step was to sprinkle sand on the wet deck. Then each seaman pushed a heavy block of stone back and forth until the deck was gleaming white. Since both the *Joseph Conrad* and the *Tusitala* had wooden decks, this task was not enjoyed by any trainee. The process was so named because, unlike in this photograph, a sailor knelt in prayer-like fashion pushing the stone forward and backward. (NARA.)

Canvas work required trainees to master four basic stitches in order to repair ripped canvas: the herringbone stitch, which had two movements; the round stitch; the flat stitch; and the baseball stitch. It also forced them to understand when to use a flat seam or a round seam. (LOC.)

As part of the preliminary training program, seamen apprentices learned the basic duties of the deck, radio, steward, and engine departments. According to the *Training Manual*, "The ordinary seaman is the junior member of the deck force. He is an all-round worker and his job is the only one open to beginners. The 'Ordinary' washes paint work, chips, scrapes, paints, scrubs the decks and performs all the various duties necessary for maintenance of a vessel in a clean and workable condition." (USMS.)

The training materials explained, "In overhauling and the repair of deck gear, cargo gear, boat gear and in other such operations, the ordinary helps the Able Seaman in many ways. It is this knowledge that he gains while doing the practical work that prepares him for advancement into the specialized field of Able Seaman." (USMS.)

Made of steel or bronzing and weighing about 200 pounds, the Lyle gun became familiar to trainees. It was designed to throw rope and other projectiles weighing up to 18 pounds as far as 700 yards from the ship to assist in any type of sea rescue. (USF.)

With Commander Krastin watching, these trainees prepare a line-throwing gun in order to practice establishing a breecher's buoy—a lifesaving device that worked from one ship to another or from the ship to shore. It consisted of a small line shot aboard a disabled ship by a line-throwing gun like the Lyle gun. Seamen from the disabled vessel were then hauled to safety using this apparatus. (USF, DSi.)

Before experiencing the breecher's buoy from the ship, trainees experienced it on land. Hanging between two masts, they were reminded to check the block throughout the entire rescue to make sure that it always ran free. (DS.)

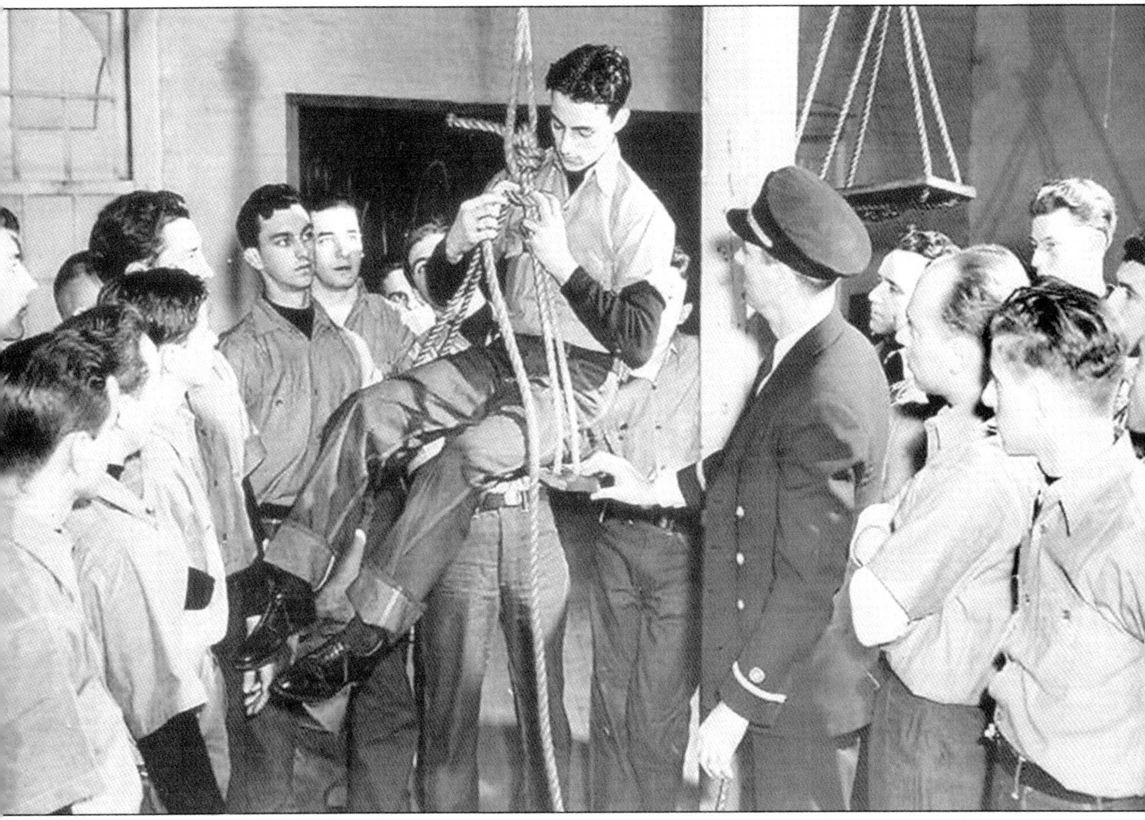

The "bos'n's chair" was a standard piece of gear on all ships used by the deck personnel. Trainees learned how to maneuver the chair, which was used for painting cargo booms aloft, handling rigging at high levels, or for lowering a man into a hold. (LOC.)

Using the international code flags and pennants, trainees received an understanding of the traditional form of ship-to-ship communications. The semaphore flag signaling system is an alphabet signaling system based on the waving of a pair of hand-held flags in a particular pattern. (NARA.)

Practicing the semaphore lessons from the classroom, trainees quickly remembered that in the first circle, the letters A to C are made with the right arm, and E to G with the left, and D with either as convenient. In the second circle, the right arm is kept still at the letter A position, and the left arm makes the movements; similarly in the remaining circles, the right arm remains fixed while the left arm moves. The arms are kept straight when changing from one position to another. (NARA.)

Communicating the letter C, the left side of trainees raise their right arm at an angle with their left arm crossing the midsection of their bodies, with the flag covering the groin area. The opposite line responds with the letter O, formed by lifting the right arm at an angle and extending the left arm across the chest area, with the flag falling just below the armpit. (NARA.)

In over a month's time, these trainees developed a working knowledge of the use of hand tools by spending 10 hours a week in the base engineering shop with Chief Fass. (USF.)

According to a maritime service training brochure, "the dexterity in the use of hand tools is essential to repair work and might well make the difference between a vessel continuing in its convoy or dropping out and becoming an easy prey for attack." (USF, USMS.)

The modern form of communication between ships, the radio required that seamen be familiar with its operation and basic components. Like the other parts of their training, merchant seamen learned this information first in the classroom and then applied it on a training vessel. (LOC.)

As the United States entered World War II, the facilities at the training station were not large enough to handle the number of men going through the programs. As a result, the maritime service made arrangements with the City of St. Petersburg to hold daily classes in a local park near the base. (USF.)

As part of the navigation instruction, five rotating platforms were installed on the west side of the base between the Administration Building and the boat dock. With trainees standing at a basic helm, the instructors stood on the opposite end turning a wheel back and forth to simulate the rocking of a ship. (DS.)

The use of the ship's compass was vital not only to the able seamen and officers who steered its course but to the survival of all crew members once inside a lifeboat. Using the compass on the *Tusitala*, trainees reviewed lessons taught earlier in the day. (GB.)

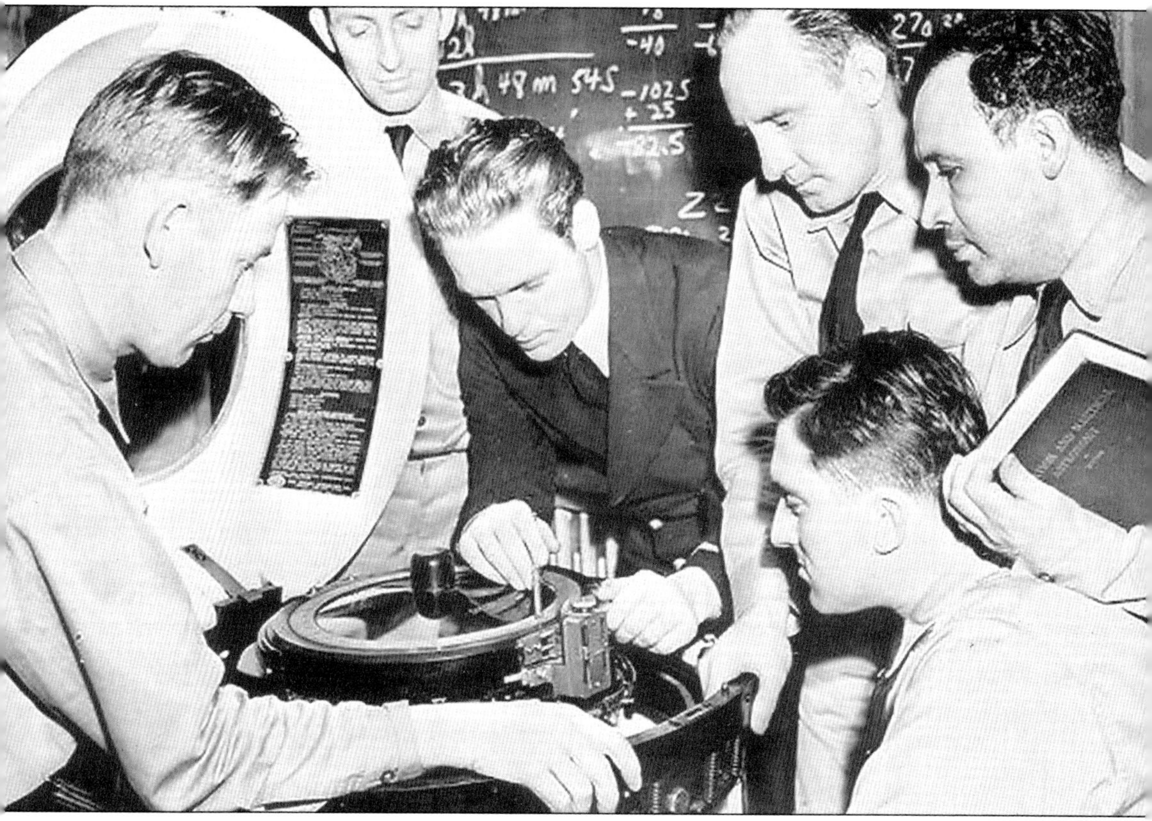

Trainees also learned how the gyrocompass operated. Using fast spinning wheels and friction forces in order to take advantage of the Earth's rotation, the compass finds true north without interference from the ship's hull, as was the case with magnetic compasses. (LOC.)

Located on the eastern seawall of the training station were seven gunnery platforms. The inclusion of gunnery training evolved from a directive requiring a specific number of merchant seamen to assist the Navy Armed Guard with the loading of ammunitions while a ship was under hostile attack. The rationale behind this directive was to reduce the number of naval personnel on board civilian ships. (DS.)

Training documents explained, "The loading machine used for training purposes simulates a gun. The actual operation of the loading crew is the same as though they were standing behind one of the latest types of guns. The same commands used on shipboard are given." (NARA.)

"War has added another 'must' to the training of seamen, and that is the proper functioning of a good gun crew," reads the *United States Maritime Service Training Manual*. "Although they may be called upon to fill a gun in action, it is essential that each man leaving the Maritime Service for sea duty be equipped with knowledge of large guns, in case an emergency arises when they are called upon to defend their cargo and their lives." (NARA.)

Four

Department Training

After two months of preliminary training, each apprentice selected the department in which he desired training, and for the next 6 to 12 weeks, the seaman was schooled for the department of his choice. As the chart indicated, the trainee could select the career path for the deck, radio, stewards, or radio departments. (USMS.)

During this second phase of training, seamen apprentices transferred to a training ship like the *American Seaman*. This ship was a steam vessel of 7,000 gross tons, 410 feet long, with a 54-foot beam with accommodations for 320 apprentices and a regular complement of 15 officers and 60 men. George Bark was on this training ship when Pearl Harbor was attacked on December 7, 1941. (GB.)

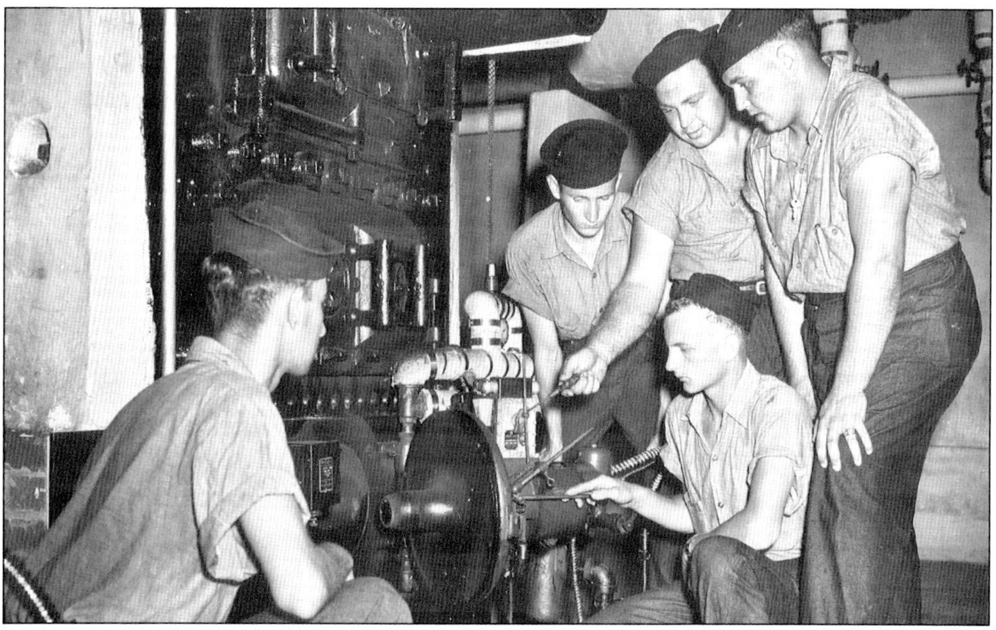

With triple-expansion steam engines and three single-ended broilers, the *American Seaman* acted as a classroom where those apprentices who selected the engine department got hands-on experience. The engine branch training was administered over a three- to eight-week period. (USF.)

Veteran merchant marine officers and men imparted their knowledge and skills to the young trainees. With ships being sunk daily, the trainees understood that their vessel could not afford to drop out of a convoy because of mechanical breakdowns. (USF.)

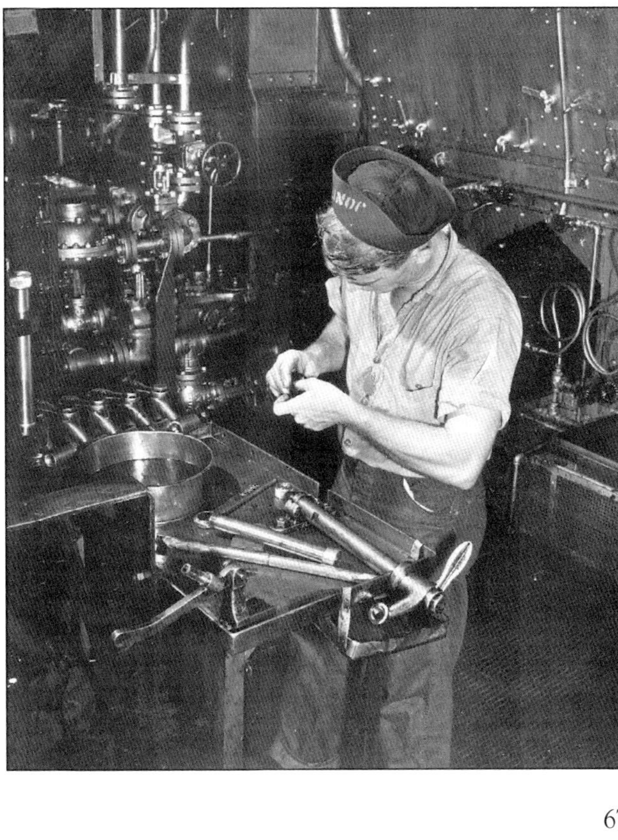

Trainees who selected the engine department stood watch on the training vessel *Vigil* as firemen, oilers, and watertenders. Nicknamed the "black gang" because of the oil on their hands and shirts, these young seamen would eventually be responsible for the safety of their ship. (USF.)

Practical training on the generator flat in the engine room of the *Vigil* taught the procedures in starting generators and shifting the load from one generator to another. (USF.)

Before the practical training came, the theoretical lessons covering all aspects of a ship's power plant facilities had to be learned. (USF.)

During classroom instruction, deck department trainees learned that the stowage of ship's cargo was based on the same principles of packing a box or trunk for shipment. The idea was to get as much in the box as possible and to pack it in such a way that the items were not damaged in transit. (NARA.)

Deck apprentices quickly realized that cargo plans differed according to the build of a ship and that the greatest threats to moving items from the shore to one of the ship's hatches were the selection of a wrong type of sling or careless winch work. (LOC.)

Once classroom instruction and model practice was completed, these apprentices received practical instruction in handling various types of cargo and winches, tending hatches, and proper stowage. (LOC.)

Deck trainees also received significant training regarding how to identify ships and planes from afar. On board the *American Seaman*, these deck trainees applied classroom lessons to the reality of working on the ship's bridge. (LOC.)

Adm. Emory S. Land, who served as the chairman of the U.S. Maritime Service and administrator of the War Shipping Administration, wrote the comments in the first page of the training manual for those who pursued the steward's specialty: "American merchant seamen are the best-fed in the world. They deserve the bountiful table set before them. That is why no effort is spared to provide the WSA [War Shipping Administration] vessels with the finest food available. We ask that it be handled in a manner befitting your great responsibility to the health, welfare and morale of those who man our Merchant Fleet. . . . Yours is an important job. Give it your best." (USMS.)

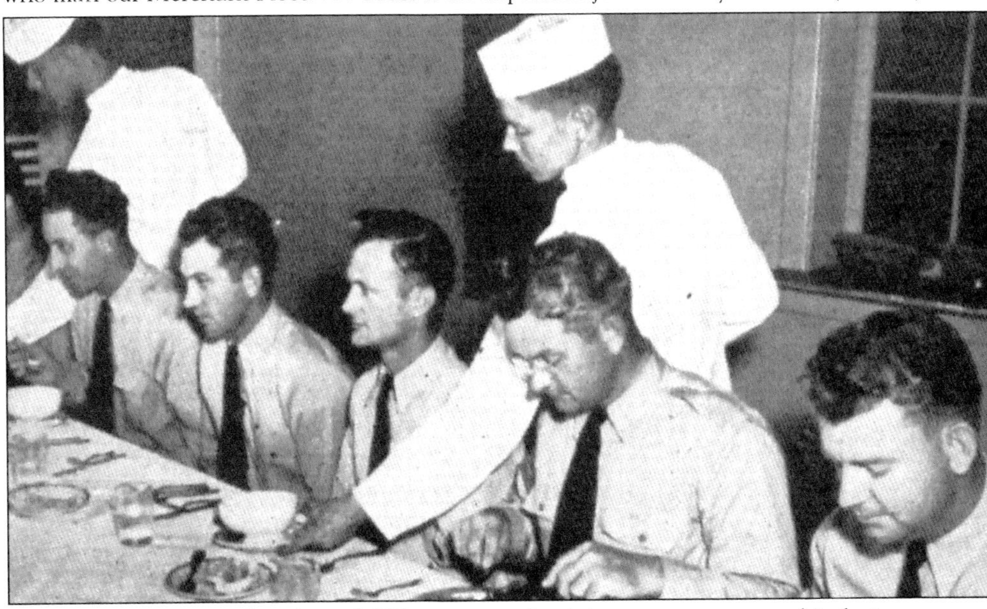

As the *Training Manual* explained, "The messman's job is not an easy one—his duties are many and varied. He is a coffeeman, assistant cook, pantry man, waiter, dishwasher, bedroom steward and porter. The messman's personal cleanliness is of great importance. The tidy messman has a healthy body, short fingernails, clean hands and neat clothing. When serving tables, he does not use strong smelling hair tonics since they may affect a person's appetite." (USMS.)

According to the War Shipping Administration's "Better Stewarding": "Tables should be washed daily with strong soap and water. Silverware should be washed in hot soapy water, rinsed after every meal and polished once a week with a silver polish paste. . . . The size of the plates vary in different sets; dinner plate is nine to 10 inches in diameter; fish and dessert plates seven to eight inches; salad plates seven inches; bread and butter plates six inches." (USMS.)

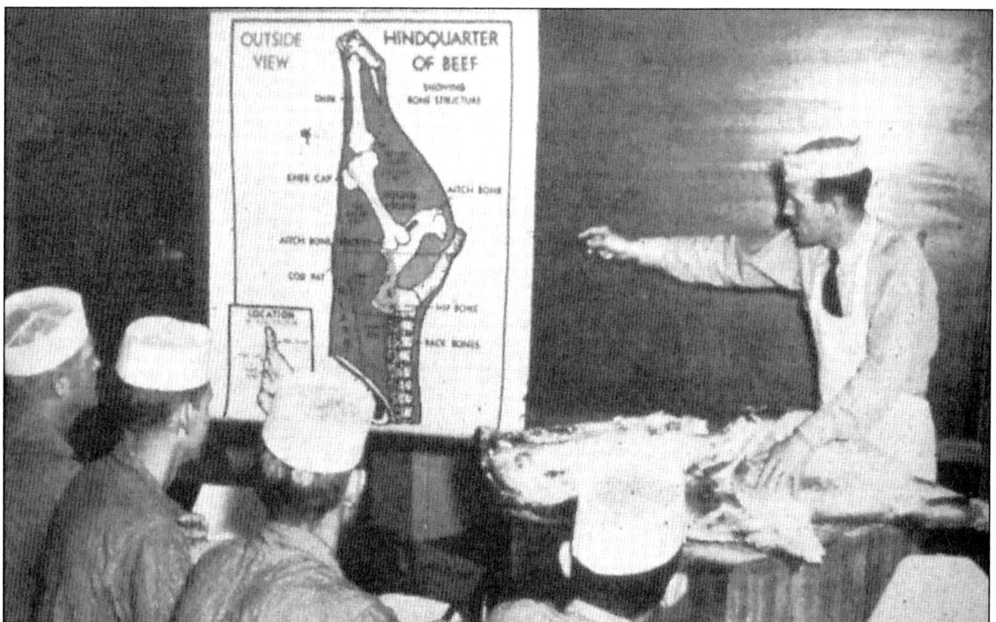

Meat was an expensive and perishable product and therefore much emphasis was placed upon its care and storage. Butchering instruction taught proper methods of cutting, boning, and tying meats, as well as correct selection of cuts of meat and their respective uses. (USF.)

According to the butchering training materials, the "WSA beef carcasses are 'fabricated' or cut into nine major sections. Fabrication eliminates waste (bones, gristle, etc.) and makes preparation and handling easier. On a poundage basis, you get the same percentage of tender cuts in a 'fabricated' carcass as in a whole one." (USF.)

NO, YOU CAN'T SERVE STEAKS AND ROASTS EVERYDAY, BUT—

There Are Plenty of Tasty Meals You Can Make with Chucks, Plates, Briskets & Ground Beef—If You Know How to Use These Beef Cuts!

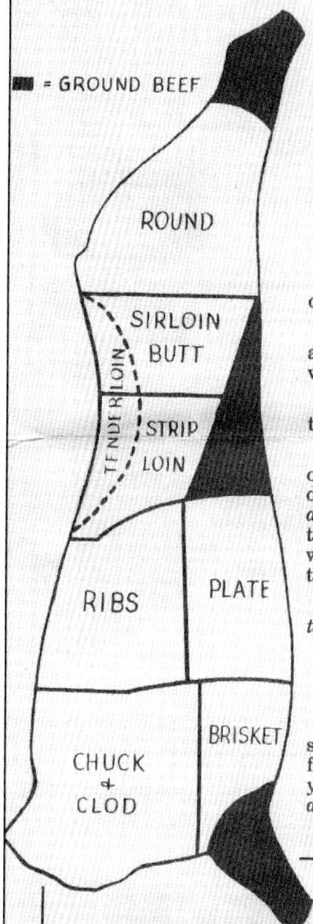

Sure, there's a *limit* to the amount of steaks and roasts you can serve.

But there *isn't* any limit to the amount of *good meals* you can make with the less-tender cuts of beef.

They can be just as good and tasty, too—*if you know HOW to use them!*

It doesn't take much skill to make ordinary hamburgers or stews—but it does take *real imagination* and *cooking ability* to create meals that won't be tiresome or tasteless . . . meals that won't be the "same old thing'—from the less-tender cuts of beef.

And YOU can create these *better-tasting* beef meals *if you'll only try!*

HERE'S HOW

There's a long list of tasty menu suggestions on the inside of this folder. Read it and find out how you can get *more variety* and *appetite appeal* from ALL your beef cuts.

If you've got real pride in your job . . . if you want to serve the *best meals* you can . . . if you want to make your work *easier*—you'll start using these beef menu suggestions today!

HERE'S WHY

There's only so much steak and roast meat on a steer—actually *85 lbs. of steaks and 70 lbs. of roasts* on an average 500 lb. "fabricated" carcass. That leaves *345 lbs. of other cuts* to be used *along with the steaks and roasts*, because the Merchant Marine buys—*and must eat*—beef by the *whole carcass.*

Unless you utilize your tender beef cuts *wisely* during the first part of a voyage, you'll find you'll have nothing left for the homeward voyage—but less-tender cuts.

What's more, you'll be leaving a big *overbalance* of these cuts for the *next* voyage. Give the men on the next voyage a break—don't use more than *your share* of roasts or steak!

WHY MERCHANT MARINE BEEF IS "FABRICATED"

To save storage space, WSA beef carcasses are "fabricated" or cut into nine major sections shown above—and are inspected, weighed, packaged under Government supervision. "Fabrication" eliminates waste (bones, gristle, etc.) and makes preparation and handling *easier*.

On a poundage basis, you get the *same percentage* of tender cuts in a "fabricated" carcass as in a whole one.

Sirloin Strip	6%	26%
Sirloin Butt	7%	
Tenderloin	3%	
Oven Prepared Ribs	10%	

Round	*22%	74%
Chuck and Clod	27%	
Plate and Navel	8%	
Brisket	5%	
Ground Beef	12%	
	100%	

* half may be used for steaks, roasts

Since you have more than *twice as much* of the less-tender cuts, you can serve steaks and roasts only once every *third* time you serve beef DON'T SERVE MORE OFTEN.

As the poster declared, cooks should not serve steaks and roasts every day. So the alternatives were explained: "A steer only has so much steak on him—like he has only so many legs—and since the Merchant Marine buys beef by the whole carcass (the way the Army and Navy do) you've got to take the other meat cuts with the steaks. As a matter of fact, out of every 500 pounds of beef aboard ship, only about 85 pounds are steak meat. The rest are what you see in the picture—round, brisket, plate, chuck, etc. So if there are plenty of stews, pot roasts and hamburger steaks on the menu, don't think the Steward is holding back the steak on you. He isn't, the steer is." (USF.)

Vegetables were an important feature of the daily menu and used more than any other food in a well-balanced diet. In addition to the varied methods of food preparation, steward trainees received instruction in the storage of vegetables. (USF.)

Baking represented one-third of cooks and bakers training. Trainees learned how to bake a variety of bread, rolls, and coffee rings. They also mastered sweet dough, butterhorns, quick breads, pies, cakes, icings, pastry cookies, and puddings. (LOC.)

The cooks and bakers instructor conducted a rigid inspection of the breads, cookies, and pastries, pointing out any defects and thoroughly explaining the methods for overcoming any faults. (USF.)

Cooking instruction for large groups required practice and supervision. Cooks and bakers prepared meals not only on the training vessels but also in the two mess halls at the training station. (USF.)

All modern cooking appliances found aboard ships in the 1940s assisted in skillfully and rapidly preparing the daily menu. Trainees were thoroughly instructed in their operation, care, and maintenance. (LOC.)

Serving up warm pies, Charles Swinkey of San Antonio, Texas, attended the Cooks and Bakers School at the maritime school in St. Petersburg and then worked on the SS *Patrick Henry* toward the end of World War II. (NARA.)

Five

Training Vessels

From the time that the St. Petersburg Maritime Service Training Station was commissioned in November 1939 until it was decommissioned in the spring of 1950, over 25,000 men trained on some combination of the *Joseph Conrad*, *Vema*, *Tusitala*, *American Sailor*, *American Seaman*, *American Mariner*, *Empire State*, and the *Vigil*. (KF.)

When the *Joseph Conrad* left New York on November 2, 1939, the ship was well known within sailing circles. Under her previous owner, George Hartford, the *Conrad* won each leg of a square-rigged ship race from the United States to Bermuda against the *Seven Seas*. (GB.)

"When I was a young kid before St. Petersburg, I remember seeing it [the *Joseph Conrad*] when it was anchored at Bayridge just outside of Brooklyn," said George Bark. The *Conrad* was the smallest mystic full-rigger ever built. Several years after riding his bike to see the ship, seaman's apprentice George Bark climbed to the highest point on ship and took this photograph using a box camera that he carefully packed in his pants. (GBi.)

In 1943, the *Conrad* was commanded by Capt. Robert Shuford. He was known for being a master sailor who understood the realities faced by his trainees once they left St. Petersburg. As a result, the *Conrad* had midnight lifeboat drills signaling all hands to abandon ship—28 men in each boat with only two left on deck. (LOC.)

Captain Shuford timed the launchings, and if the process took too long, he waited 20 minutes until the seaman on watch went back to bed and then gave the abandon ship order again. In the *Maritime Service News*, Captain Shuford explained, "They don't like that, but they know it is the kind of training on which their lives will depend if they get torpedoed." (USMS.)

Designed in 1923 for financier E. F. Hutton, the training vessel *Vema* was originally christened the *Hussar*, the fourth of a series of five ships. The sailing ship was the most extravagant privately owned ship of her time. (DS.)

Sold to George and Maude Vettlesen in 1930, the ship was renamed the *Vema*. The 182-foot steel hull was commissioned by the Coast Guard in July 1941 and was initially utilized at a training station in New York. (DS.)

In 1953, Columbia University purchased *Vema* from the Maritime Commission and converted it into a marine survey vessel. In 1982, Capt. Mike Burke, founder of the Windjammer Barefoot Cruises, purchased the sailing ship and refitted it to its former glory. Today the vessel sails under the name *Mandalay* and travels 500 miles from Antigua to Grenada carrying accommodations for 72 passengers and 28 crew. (DS.)

On May 14, 1940, the *Tusitala* was given her final inspection before leaving Tompkinsville, Staten Island, to be towed 1,736 miles to St. Petersburg. (DS.)

The federal government spent $31,000 in refurbishing the 261-foot vessel. The *Tusitala*'s pine deck houses were burnt off and replaced with steel decks and training rooms. There were electric cookers, outlets for electric razors, cooper screens, and expanding wire mesh over all lights and windows to keep out the mosquitoes. (GB.)

Built in 1883 as the *Inveruglas*, the ship was the last built by the Steel of Greencock Scotland. The vessel then sailed to Australia where it was renamed the *Sierra Lucana*. From 1904 to 1923, its Norwegian owners called it the *Sophie*. (AV.)

The last full-rigged ship to be used as a commercial cargo transport from Honolulu to the continental United States, the *Tusitala*, became a stationary training ship in St. Petersburg. With double-decked beds for 120 seamen and 30 officers, George Bark, pictured here, was one of the first trainees to sleep and attend classes on the vessel. (GB.)

The *American Seaman* was the first ship to arrive in St. Petersburg in November 1939. The 10,000-ton converted merchant ship was capable of carrying 300 trainees plus officers and crew. (AV.)

Making periodic cruises of two to three weeks to numerous countries in the northern hemisphere, trainees received actual sea experience and were able to apply lessons learned in the classroom and on land to reality. Seen above, the *American Seaman* passes through the channel in front of the Morro Castle in Havana, Cuba. (LOC.)

While on a training cruise with the *American Seaman*, George Bark played the accordion for friends Nick Bertuccio (second from left), George Boman (on the far right), and Herman Wirmecki (sitting above). The average cruise left out of St. Petersburg and, four days later, arrived in Kingston, Jamaica. After two days, the ship departed for Port au Prince, Haiti, where it remained for an additional three days. From Haiti, the *American Seaman* sailed to Havana, Cuba, and then returned to St. Petersburg. (GB.)

Already over 20 years old when transferred to the maritime service in 1939, the *American Seaman*, a Liberty ship built in 1919, was properly maintained thanks to the Maritime Commission. Every nine months, maritime service training ships used in tropical waters were brought to dry-dock for cleaning. The first step was to scrape the hull, ridding the ship of all debris from the side and bottom. Sometimes as much as 30 tons of barnacles were removed. The second step was to hose down the hull with fresh water. All sea valves and sea connections were opened, examined, and cleaned. Great lengths were taken to check for leaks and any marks of wear and tear. Then the outside of the ship was painted with two coats. The initial coat was an anti-corrosive paint, followed by a second of anti-fouling paint. What was amazing about the *American Seaman*'s dry-dock in 1945 was that all of this, as well as the repairs on the interior, was completed in 19 hours. (GB.)

Converted into a training ship by Bethlehem Steel Company in Baltimore, the *American Sailor* was approximately 409 feet long with steam turbines and a single propeller. Prior to its acquisition by the Coast Guard, the ship was called the SS *Edgemont*. (DS.)

In November 1939, the Coast Guard was transferred to the Navy Department in preparation for the pending war in Europe. Upon news of the attack at Pearl Harbor, all maritime training ships that entered international waters were painted battleship gray. In 1953, the *American Sailor* was decommissioned and sold for scrap metal. (DS.)

The *American Mariner* was, as Herb Hahn wrote, "a companion to two elderly cargo carriers (*American Seaman* and *American Sailor*) who had entered the teaching profession." With a three-cylinder reciprocating steam engine, two boilers, and one propeller, the Liberty ship was 441 feet long and was considered by most to be the best training ship in the Maritime Commission's fleet. (DS.)

The *American Mariner*'s name was originally *George Calvert*, but Adm. Telfair Knight suggested a more appropriate title given that it was the first ship designed and built for the training of American merchant seamen. So on December 30, 1941, the ship's name was officially changed to *American Mariner*. (DS.)

As the realities of postwar America came into focus, funding for the maritime service began to dwindle. As a result, the *American Mariner* was reassigned to the Massachusetts Maritime Academy in 1948. (DS.)

As late as 1990, the *American Mariner* was still in service as part of a training program. With its hull firmly on the floor of the Chesapeake Bay and three compartments flooded to the new water line, the ship served as a target vessel for the Naval Air Test Center (NATC). It has also been used in radar tests and for magnetic anomaly detection exercises. (Courtesy of NATC Patuxent River, Maryland.)

Before being assigned to St. Petersburg, *Empire State* served as a training ship for the New York State Merchant Marine Academy. Built in 1919 as the *Shaume*, the ship served as the flag ship of the Navy's Pacific Fleet Supply Train from 1923–1930. (DS.)

The training ship *Empire State II*, built in 1944 was considered the perfect replacement for the *Empire State I* with twin screws, two twin engine rooms, and turbo-electric power. (DS.)

The training vessel *Vigil* housed the preliminary training programs for the engine and deck apprentices. The ship arrived at the training station in 1942. (AV.)

Upon news that the St. Petersburg training station was to be decommissioned, the *Vigil* was stripped of all equipment and towed to Mobile Harbor, Alabama, where it was to be scrapped. (DS.)

There were two tugboats, the *Tickfaw* and the *Morganza*, assigned to the training station to assist with the position of ships in Bayboro Harbor as well as for training coal-burning firemen. The *Morganza*'s mascot was a small calico cat. (GB.)

Another vessel at the training station was the *Storm King*. It was used for training exercises as well as for recreation time for the officers and their guests. (DS.)

Four weeks on board a ship gave trainees the opportunities to put into practice the basic courses received at the station. Upon returning to shore, trainees were awarded their Seaman's Certificate and placed in the U.S. Maritime Service pool from which future assignments were made. (GB.)

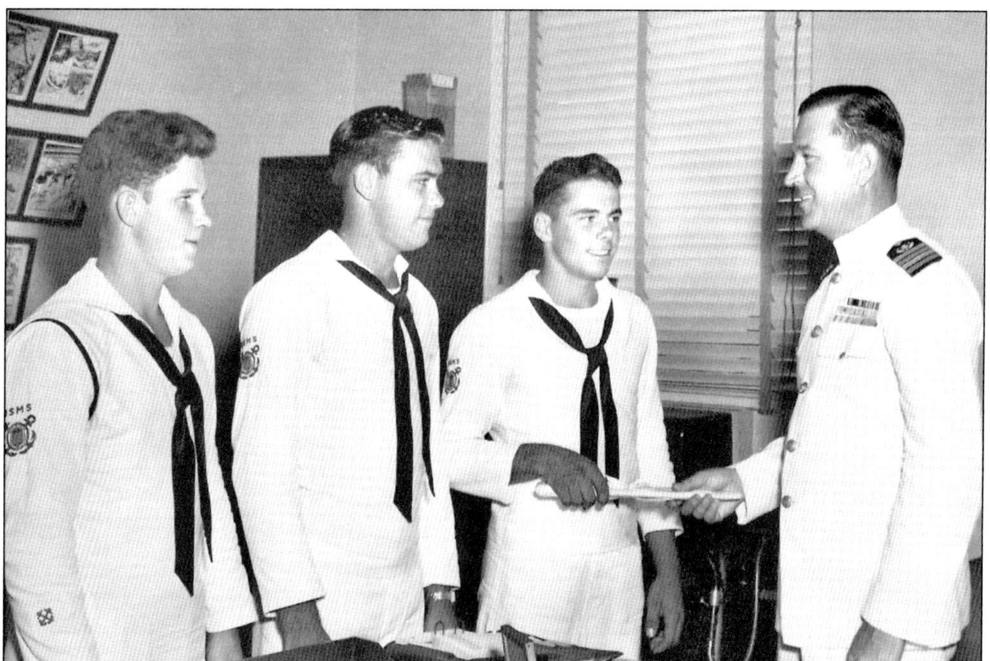

Students who finished their training with the highest grades were awarded certificates by the commanding officer. These trainees were congratulated by Capt. H. S. Tiedemann, who later became an admiral as well as the superintendent at the U.S. Merchant Marine Academy. (DS.)

Six

Officers and Personnel

The officers of the St. Petersburg Maritime Service Training Station came from all around the country and from diverse backgrounds. The base consisted of the following departments: training, finance and supply, maintenance, ship's service, commissary, medical, and personnel. (USF.)

Comdr. H. J. Tiedemann, seen here, was in charge of the St. Petersburg facility twice during its history. During his first assignment in 1943, Tiedemann was part of the Naval Reserves; during his second tour from 1946 to 1950, he was promoted to captain and served in the maritime service. The first superintendent of the training station, Comdr. W. W. Kenner of the U.S. Coast Guard, held the position for three years. Comdr. P. C. Mahady was responsible for the training station for 12 months from 1942 to 1943. Captains G. E. Harrington, E. F. Jaeger, and J. H. Masse acted as superintendents from 1944 through 1946 respectively. (DS.)

The second-in-command at a training station according to the maritime service's organization chart was the executive officer. He acted as the principal advisor to the superintendent. He also directed the general day-to-day operation of the station and acted as the commanding officer during the superintendent's absence. Seen above congratulating several honor graduates, Comdr. David McMichael was the executive officer under Captain Tiedemann and a few years later became the captain of the NS *Savannah*, the first of three nuclear-powered cargo ships ever built. (DS.)

When the base was decommissioned in 1950, the training department was under the direction of Lt. Comdr. E. G. Hammond. Mrs. M. E. Minor was his administrative assistant. The department had 15 officers and 15 non-commissioned officers. (USF.)

Lieutenant (junior grade) Keuther headed the cooks and bakers division of the training department. The St. Petersburg training station became known for the quality personnel it trained in this program. (DS.)

Lt. Comdr. C. R. Lundstrom headed up the finance and supply department. Together with three additional officers, five non-commissioned officers, and Mrs. J. G. Gray, the civilian administrative personnel, these individuals ensured that materials needed for the operations of the base were ordered and that pay was distributed to all personnel. (USF.)

Under the watchful eye of Lt. J. C. Berry, the maintenance department was responsible for ensuring that any mechanical breakdowns affecting the day-to-day operations of the seven-acre base were quickly handled. (USF.)

Lt. T. J. Shannon headed the ship's service department when the base was decommissioned. Similar to today's base exchange, Shannon's team ensured that trainees and personnel were able to buy toiletries and other necessities when on the installation. (USF.)

Located in the Administration Building was a magnificent library with over 2,000 books. The responsibility for maintaining the library and adding to its collection was assigned to Lt. Thomas P. Webster, a local St. Petersburg resident who joined the maritime service. (DS.)

Lieutenant Webster's daughter, Mildred, was the wife of Florida's U.S. senator, Claude Pepper, seen here at center with Captain Mahady in the summer of 1942. After meeting in 1930, the future senator and Mrs. Pepper courted for six years and married in St. Petersburg on December 29, 1936. Senator Pepper was a frequent visitor to the training station. During his visits, boat races between training crews were held, and his speeches were broadcast on WSUN. (USF.)

Denis Quilligan (left) and Donald Saxer worked at the training station together until 1946. Originally from Granite City, Illinois, Quilligan joined the maritime service in 1942 after being turned down by the armed services due to his color blindness. After completing his training in St. Petersburg, he sailed in the North Atlantic and the Mediterranean, never having a ship blown out from beneath him but experiencing the terrors of a ship being attacked. He turned to the St. Petersburg station as a quartermaster in 1944, and upon leaving the maritime service, he joined the St. Petersburg Police Department. (DS, DQi.)

The personnel department was responsible for the administration of all paperwork on base including local enrollment, sick leave, personal leave, disenrollment, and personal movement from ship to shore or base to base. (DS.)

Throughout the base's history, civilian women were employed to assist with the numerous administrative tasks. Ironically, inside the Administration Building, women were required to use the enlisted men's restrooms because designed bathroom facilities were not included for women. (USF.)

In January 1944, Ens. Charles Keller reported for duty at USMSTS St. Petersburg as a physical fitness officer. Six months earlier as a New York Yankee, he was selected to the American League All-Star team. During the 1943 Major League Baseball season, Keller led the league with 106 walks and had 31 home runs. His offensive talents in the 1943 season ranked him highest among players when adding together his on-base percentage and slugging percentage. He was one of 18 New York Yankees to leave the ballpark during World War II. At the end of war, Keller returned to the Yankees, where he played for an additional four-and-a-half years before being traded to the Detroit Tigers. In 1952, Keller returned to the pinstripes and left baseball as a New York Yankee. Most avid baseball fans from this time remember Keller most for his performance in the 1939 World Series against the Cincinnati Reds, where he became the first rookie to hit two home runs in a World Series game. After baseball, he started a very successful horse-breeding farm in Maryland, which is still managed by his family today. (MK.)

Dental technician Lloyd Green left the maritime service and went to work for the Veterans Administration in the New York area. (DS.)

Security Officer Joe Hitchcock (left) was actually part of the U.S. Coast Guard. The individual behind him was from the U.S. Naval Reserve. (DS.)

Standing in front of the cargo training center, Master-at-Arms Max Flagel headed up the security detail at the training station. (DS.)

Comdr. D. S. Goddard was the commanding officer of the U.S. Maritime Service training ship *American Mariner* from 1948 to 1950. It was on his watch that the *Mariner* departed St. Petersburg for her new assignment in Massachusetts. (DS.)

Commander Goddard's executive officer from 1948 to 1950 was Lt. Comdr. Henry G. Simmons. (DS.)

An officer on board the *American Seaman* uses the sextant to determine the ship's navigation. By looking through the eye piece, the officer finds the horizon and then adjusts a moveable arm with a second mirror to the position where the sun is reflected upon it. What resulted was the sun being superimposed on the horizon, and the angle between the two is read off to the navigator. This angle and the time when it was taken were then used to calculate the ship's position on a nautical chart. (LOC.)

These officers from the *American Seaman* were highly respected by the trainees because of their seamanship and keen knowledge of the dangers of maritime work, especially in a time of war. (GB.)

During the preliminary training, marching drills were considered an important part of the transition from a civilian to maritime mind-set. The drills reiterated the importance of following directions, maintaining concentration, and working as team. (USF.)

Part of the daily routine for trainees was the muster before classes began. This was a brief opportunity for the base leadership to monitor the military-like discipline of the training program. (USF.)

Formal inspections on ship required preparation by trainees and staff. The captain's party reviewed every aspect of the ship. Trainees stood at attention while the captain made his review. If inspections were passed, extra liberty was given for the job well done. (NARA.)

Following in naval tradition, the St. Petersburg training station had its own chaplain. According to base documents, Lt. J. A. McKenna was the base's Catholic chaplain from 1943 through the summer of 1944. However, the chaplain shown in this photograph on the far left is Captain Thomas from the navy's chaplain corps, which was established in 1775. The Reverend Benjamin Balch, a Harvard graduate, was commissioned as the first navy chaplain on October 30, 1799. (DS.)

On a wet January morning standing in the wheelhouse of the *American Mariner*, Captain Tiedemann (left) welcomed the superintendent of the Maritime Academy in Castine, Maine, Rear Adm. W. W. Warlick (far right). Capt. G. A. Coas (center), commanding officer of the *Mariner*, hosted the admiral, his officers, and a delegation of midshipmen during their visit to the station. (USF.)

Seven

The Base and Its City

What was obvious from the beginning was St. Petersburg's love for the maritime service training station. Although it was not always a perfect relationship, the maritime personnel and trainees became not just visitors to the city but part of it. Aaron Varn, who grew up across the bay, had this picture taken at the Million Dollar Pier with his girlfriend, Cecile Cooper, in 1941 while on leave from the training station. The couple was married two years later and still resides in the Tampa Bay area. (AV.)

Until the base received a chaplain, liberty was given to trainees who decided to attend worship services in town. Trainee George Bark had this picture taken outside St. Mary's Catholic Church. (GB.)

Trainee Kenneth Fountain recalled taking flying lessons at Albert Whitted Municipal Airport, located just outside the base's perimeter. "I already had a few hours flying instruction in my home, so decided to get a few more. An old aviator by the name of Matt Pelling ran a small flying school there at the airport. His instructor was named John Benefield. I did get to log a few more hours and managed to solo shortly after my 18th birthday." (USF, KFi.)

Started in 1913 by Arthur Johnson, the St. Petersburg Festival of States began merely as a fall festival. By the time this photograph was taken in 1949, the fall festival had turned into a week-long celebration with over 400 marching band competition held through the city. The maritime service training station participated in the parade each year until the base was decommissioned. (DS.)

The officers and trainees from the base also participated in the Armistice Day celebrations commemorating the end of World War I on November 11, 1918. The holiday was renamed Veterans' Day in 1952 to honor those who served in all American wars. During the 1947 Armistice parade, Captain Tiedemann (far left) rode with an unidentified army officer and a local official. (USF.)

Another holiday celebrated by the city and the base alike was Maritime Day, May 22. The 2003 National Maritime Day Proclamation read, "Today, as in the past, America depends on our maritime services to help ensure our security, promote our prosperity, and advance the universal hope of freedom. We honor the service and proud history of our merchant mariners and also recognize their important contributions in strengthening our economy." (USF.)

On the training installation, Maritime Day was marked with a 21-gun salute, the ringing of the bell eight times for mariners on eternal voyages, and the laying of a wreath off the most southern point of the base. The 2003 proclamation from Pres. George Bush reminded Americans: "Even before the United States declared war, merchant mariners were making perilous runs to Europe with desperately needed supplies. President Franklin Roosevelt, the first President to issue a proclamation honoring merchant mariners, wrote of their role during wartime: 'They have delivered the goods when and where needed in every theater of operations and across every ocean in the biggest, the most difficult and dangerous transportation job ever undertaken.' " (USF.)

When the day was done, trainees went back to their dormitories or ships to study and relax. Others discussed how to spend their time off base. Some planned trips to Tampa to visit Ybor City's nightclubs, while others wrote letters home to parents and sweethearts. (NARA.)

Ed Kinter lucked out when it came to housing. He enrolled in the maritime service at a time when the St. Petersburg base was at capacity plus more. After processing at the base, his company marched to the Vinoy Hotel—one of the most luxurious resorts in the city prior to the Depression. Kinter and others spent six weeks sleeping, showering, and studying in this distinctive St. Petersburg landmark. The maritime service also rented out the Soreno Hotel for housing seamen apprentices, while the Colonial and Concord Hotels were used to house the Army Transportation Corps and its officer cadets. (DS.)

Trainees were also given the opportunity to take the sailing boats out for recreation as long as no civilians were aboard. Aaron Varn (right) and his buddies took this photograph just off the Million Dollar Pier built in 1923. (AV.)

Purchased by the U.S. Army in 1942, the Don Cesar Resort on St. Pete Beach turned into a convalescent home for battle-fatigued airmen. Trainees from the maritime service training station were allowed to use the beach during their weekend liberty. (GB.)

When *Action in the North Atlantic* was released in 1943, trainees marched from the station to the Playhouse to see Humphrey Bogart as Lt. Joe Rossi, the first officer on a Liberty ship. When his tanker is torpedoed by a German U-boat, Bogart takes command of a Liberty ship, and his convoy must play cat and mouse with a German wolf pack. Ironically Bogart's character is later reassigned to the *Sea Witch*, the vessel built in Tampa. (DS.)

In the film, regarded as one of the most realistic movies about the merchant seamen during World War II, the director required the German characters to speak in their native tongue and did not use subtitles. Many attributed the realism of the movie to the accurate representation of convoy battle tactics and techniques. (DS.)

Because alcohol was not allowed on the training station, many personnel parties were held at local restaurants and bars. At a local venue, bartender Jeff Taylor (left) poured another round for (front to back) Chiefs Langshaw, Trafton, Mogelnicki, Williams, Gillette, and Saxer. The men second from left and far right in the back are unidentified. (DS.)

The 1949 personnel department party was held at a restaurant in downtown St. Petersburg. Civilian workers and their dates were invited to attend. (DS.)

Eight

Leaving USMSTS St. Petersburg

As the war raged in Europe and the Pacific, the St. Petersburg Maritime Service Training Station ushered thousands of trainees through the basics, knowing what awaited them upon graduation. "Snuffy" Smith, the first man in the Air Tasking Order (ATO) 8th Air Force to receive the Congressional Medal of Honor, wrote a greeting from the U.S. Merchant Marine to the German submarines on armaments being loaded on planes. (NARA.)

Graduating with honors was marked by extra leave and a photograph with the base commander, Captain Tiedemann. (DS.)

Just before graduation, the trainees' unit stood in front of the Administration Building and took a graduation photograph. For many trainees, this is the only surviving image of their time in St. Petersburg. (NARA.)

Graduation Day was held on the southern point of the training station with the *Joseph Conrad* in full flags. Civilian guests were invited to witness the graduation, and the entire corps of trainees was present for the graduating class. (USF.)

After graduation was completed, seamen received a complete issue of clothing and other items needed for work on a ship. Officers inspected each man's issue to ensure that every item was present before packaging. (DS.)

With their final issues package and the last inspection completed, these trainees anxiously awaited the arrival of the buses. (DS.)

Graduation from the maritime service meant that trainees were given their training certificates and were ready for employment in the merchant marines. Although every trainee had the option to quit at anytime, the training station lost less than 15 percent of a class because of disenrollment. (USF.)

Leaving the training station by Florida Motor Lines, these trainees headed to Jacksonville and then to Savannah for deployment. Ed Kinter arrived in Savannah and was quickly assigned to the *George E. Badger*. (DS.)

Departing St. Petersburg by Pullman cars meant trainees were heading to New Orleans or California. Aaron Varn left St. Petersburg and traveled across the country to San Diego, where he awaited his first position on a ship. (USF.)

Unknown to most Americans, the U.S. Merchant Marine was an integrated service. African Americans and other minorities served on cargo vessels with white personnel. However, because of its segregation laws, no black enrollees trained at the St. Petersburg Maritime Service Training Station, and there were no black officers. (NARA.)

There were several African American captains who served during World War II and several ships christened for famous black Americans. This photograph was taken on February 8, 1943, just after Capt. Hugh Mulzac (fourth from the left) and the crew of a new Liberty ship, the SS *Booker T. Washington*, completed its maiden voyage to England. (LOC.)

By January 1945, over 6,000 merchant seamen had been lost at sea due to German submarine attacks. In that month alone, 65 merchant marines lost their lives in the Atlantic and Pacific campaigns. (LOC.)

With foam beneath his feet, this merchant seaman battled the flames on his ship, trying desperately to prevent the fire from growing. The heat from the fire was so intense that it twisted the iron stairs. (LOC.)

Abandoning ship was the only option for some merchant seamen. Rescued in the rough waters of the North Atlantic, these men boarded the ship by climbing up ropes similar to those used on the climbing apparatus. St. Petersburg trainee Rex Farley abandoned ship on his first voyage to sea. "At two o'clock in the morning, boom! Two torpedoes hit us," Farley explained. "And the ship went down like a rock in about five minutes. When I stepped off the ship's deck into the lifeboat, it was already floating." Farley continued, "It was a bright full moon night. We got everyone in our boat, and then got the oars and started rowing. The ship, then, went down by the stern with us no farther than fifty feet away." (RFi.)

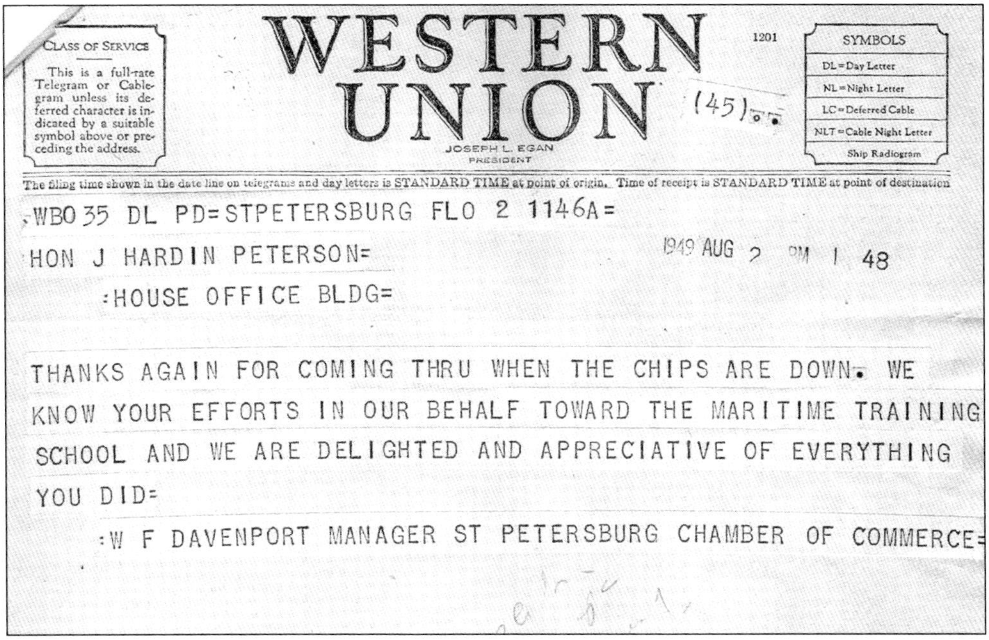
CLASS OF SERVICE

This is a full-rate Telegram or Cablegram unless its deferred character is indicated by a suitable symbol above or preceding the address.

WESTERN UNION

JOSEPH L. EGAN PRESIDENT

1201

SYMBOLS

DL=Day Letter

NL=Night Letter

LC=Deferred Cable

NLT=Cable Night Letter

Ship Radiogram

The filing time shown in the date line on telegrams and day letters is STANDARD TIME at point of origin. Time of receipt is STANDARD TIME at point of destination

WBO35 DL PD=STPETERSBURG FLO 2 1146A=

HON J HARDIN PETERSON=

1949 AUG 2 PM 1 48

:HOUSE OFFICE BLDG=

THANKS AGAIN FOR COMING THRU WHEN THE CHIPS ARE DOWN. WE KNOW YOUR EFFORTS IN OUR BEHALF TOWARD THE MARITIME TRAINING SCHOOL AND WE ARE DELIGHTED AND APPRECIATIVE OF EVERYTHING YOU DID=

:W F DAVENPORT MANAGER ST PETERSBURG CHAMBER OF COMMERCE=

In the summer of 1949, the St. Petersburg newspapers began publishing reports that the Maritime Commission was considering closing the city's training station. Earlier in the year, the station had been cut from the Maritime Commission's budget. Due in large part to the efforts of Rep. J. Hardin Peterson, the base was given $1,054,094 in funding through an amendment introduced by the congressman. By October, Peterson, along with two other members of Congress, was told by President Truman at the White House that the size of the maritime service training program had to match the number of American ships in service, which had significantly reduced since the end of the war. Local officials flooded Peterson's office with telegrams of support and encouragement for the maritime service. (UF.)

The *St. Petersburg Times* reported on February 16, 1950, that the maritime service training station had been dropped again from the Maritime Commission's budget. This time it proved fatal. The last official act logged at the station was performed at 6:49 p.m. on April 1, 1950, when the flag was lowered and furled. (USF, DSi.)

A stand-by crew was assigned to monitor the base to ensure that there were no intruders and that the facilities were adequately maintained. (DS.)

In 1954, the maritime service was discontinued, and the training of merchant seamen was limited to the Merchant Marine Academy and six state academies. The facilities built for the USMSTS St. Petersburg are still used for educational purposes today. The Administration Building is now home to the College of Marine Science of the University of South Florida St. Petersburg, whose program is nationally acclaimed. Prior to USF's presence, the building was the first home for Florida Presbyterian College. The school later moved to a new location and was renamed Eckerd College. Whether training college students, conducting marine research, or training 25,000 men for the Merchant Marine, education has always been a large part of this facility's history. (DS.)

BIBLIOGRAPHY

Cornell, Felix M. *American Merchant Seaman's Manual: For Seaman By Seamen*. New York: Cornell Maritime Press, 1942.

Hahn, Herbert P. *American Mariner: A Documentary Biography*. Kings Point: American Merchant Marine Museum Foundation, 1990.

How to Prepare and Stow Food. War Shipping Administration, 1942.

The Mast Magazine. Volume 3 (1946): 28

The Mast Magazine. Volume 3 (1946): 33

The Mast Magazine. Volume 3 (1946): 23

"Maritime Service War Service Bill." *The Mast Magazine*. Volume 2 (1945): 2–5

The Mast Magazine. Volume 2 (1945): 12, 32

The Mast Magazine. Volume 2 (1945): 26

Turpin, Edward A. *Merchant Marine Officers' Handbook*. New York: Cornell Maritime Press, 1943.

United States Maritime Service Training Manual. Cornell Maritime Press, 1943.

United States Maritime Service Training Manual. War Shipping Administration, 1942.

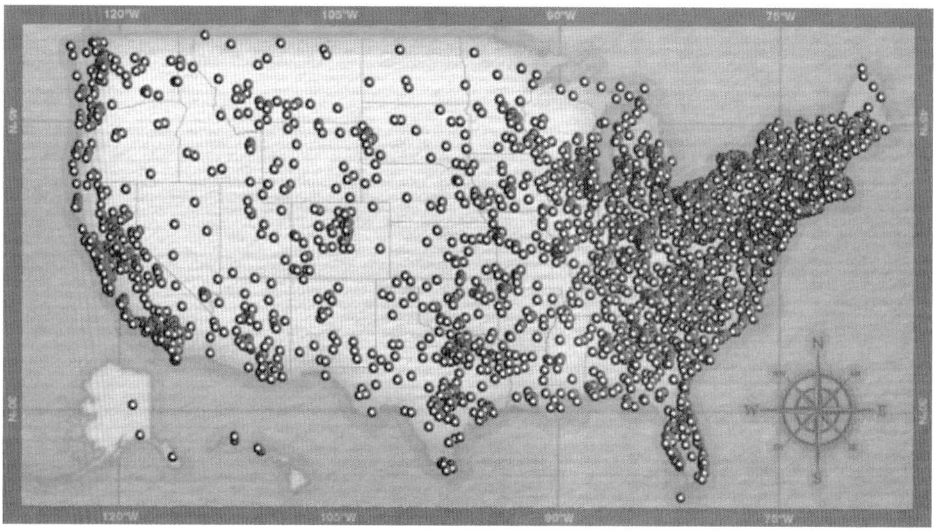